Power Systems

Electrical power has been the technological foundation of industrial societies for many years. Although the systems designed to provide and apply electrical energy have reached a high degree of maturity, unforeseen problems are constantly encountered, necessitating the design of more efficient and reliable systems based on novel technologies. The book series Power Systems is aimed at providing detailed, accurate and sound technical information about these new developments in electrical power engineering. It includes topics on power generation, storage and transmission as well as electrical machines. The monographs and advanced textbooks in this series address researchers, lecturers, industrial engineers and senior students in electrical engineering.

Power Systems is indexed in Scopus

More information about this series at http://www.springer.com/series/4622

Zhibin Qiu · Jiangjun Ruan · Shengwen Shu

Air Insulation Prediction Theory and Applications

Science Press
Beijing

Springer

Zhibin Qiu
Department of Energy and Electrical Engineering
Nanchang University
Nanchang, Jiangxi province, China

Jiangjun Ruan
School of Electrical Engineering and Automation
Wuhan University
Wuhan, Hubei province, China

Shengwen Shu
College of Electrical Engineering and Automation
Fuzhou University
Fuzhou, Fujian province, China

ISSN 1612-1287 ISSN 1860-4676 (electronic)
Power Systems
ISBN 978-981-10-5162-3 ISBN 978-981-10-5163-0 (eBook)
https://doi.org/10.1007/978-981-10-5163-0

Jointly published with Science Press, Beijing, China
The print edition is not for sale in China Mainland. Customers from China Mainland please order the print book from: Science Press.

Library of Congress Control Number: 2019935148

This Springer imprint is published by the registered company Springer Nature Singapore Pte Ltd.
The registered company address is: 152 Beach Road, #21-01/04 Gateway East, Singapore 189721, Singapore

Preface

Air is the most commonly used insulating medium for electrical equipment. Over a century, air discharge phenomena have been extensively investigated, both experimentally and theoretically. The issue to be solved for air insulation is how to determine the breakdown voltages of various gap geometries under different applied voltage types, which contributes to selecting suitable electrode structure and gap distance, so as to guide the insulation design of electrical equipment. Up to now, air discharge theory is not yet perfect, and therefore it is difficult to determine the breakdown voltage accurately by theoretical calculation. The engineering design still relies on the measured data and empirical laws obtained by discharge tests. The problem of how to predict air gap breakdown voltage has arisen great attentions by researchers for a long time.

Air discharge process is of poor controllability, measurability, and repeatability. Therefore, scientific modeling of air discharge process is very difficult since there are no governing equations to describe this complicated phenomenon. For an air gap under a given applied voltage, the discharge process varies every time even under the same test conditions, but the dispersion of the breakdown voltage is within a certain range. Therefore, under a specific applied voltage and atmospheric condition, air gap breakdown voltage is determined by the gap structure, which can be characterized by the static electric field distribution. Under switching or lightning impulse voltages, the dielectric strength of an air gap is also related to the voltage waveform. On the other hand, air discharge can be viewed as a release of the capacitive energy stored in the electric field. The static electric field distribution and applied voltage waveform respectively determine the spatial and temporal distribution characteristics of the energy storage status. If we can establish the relationship between the energy storage status before discharge initiation of an air gap and its breakdown voltage, it is possible to achieve breakdown voltage prediction without directly considering the complex and random discharge process.

We wrote this book in belief that applying artificial intelligence and machine learning algorithms to breakdown voltage prediction of air gaps would open the possibility of guiding external insulation design of transmission and transformation projects by simulations instead of experiments, so as to achieve the goal of "computational high voltage engineering (CHVE)", a new concept recently proposed aiming at solving the scientific and technical issues in high voltage engineering by simulation techniques instead of traditional experimental methods. This book proposes the air insulation prediction theory and method. Predictions of discharge voltages in different cases are discussed and worked out by simulations. The novel ideas proposed in this book help to provide references for solving bottlenecks of traditional air discharge theories. We hope that this book will be useful for graduate students, scientific research personnel, and engineering staff working in the area of air discharge theories and high voltage engineering.

This book consists of seven chapters. Chapter 1 introduces the background of air insulation prediction research. Chapter 2 presents the theoretical foundation of the proposed air insulation prediction theory. Chapter 3 introduces the air gap discharge voltage prediction model. Chapter 4 presents the corona onset voltage prediction of different electrode structures by the proposed method. Chapter 5 describes power frequency breakdown voltage prediction of short air gaps with typical and atypical electrodes. Chapter 6 deals with switching impulse and lightning impulse discharge voltage prediction of long air gaps with different geometries and under different applied voltage waveforms. Chapter 7 presents engineering applications of air insulation prediction model in discharge voltage prediction of some actual configurations.

The research work was carried out for several years. Prof. Jiangjun Ruan first proposed this research plan, and Dr. Shengwen Shu made some interesting explorations during the period of his Ph.D. studies. Then Dr. Zhibin Qiu carried on this work until now and made some fruitful progress, and he is still engaged in studies on this area. Most of the work presented in this book was fulfilled without any financial support, and partly was supported by the Open Foundation (NEL201509) of National Engineering Laboratory for Ultra High Voltage Engineering Technology (Kunming, Guangzhou) and China Postdoctoral Science Foundation (2016M 602354). The authors would like to express deep gratitude for the support.

The authors are also grateful to A. P. Daochun Huang, Dr. Feng Huo, Dr. Ziheng Pu, Dr. Chao Liu, and several postgraduate students, including Congpeng Huang, Wenjie Xu, Xuezong Wang, Qi Jin, Wei Lu, and Yongqing Deng, for their contributions, made during the period of their postgraduate studies undertaken at Wuhan University, to part of the achievements presented in this book. The authors also appreciate all colleagues and all the graduate students associated with this work in Nanchang University, Wuhan University, and Fuzhou University, for all assistance provided.

Finally, special thanks are expressed to our families and our great motherland.

Nanchang, China — Dr. Zhibin Qiu, Nanchang University

Wuhan, China — Prof. Jiangjun Ruan, Wuhan University

Fuzhou, China — Dr. Shengwen Shu, Fuzhou University

Contents

1 Background of Air Insulation Prediction Research 1
1.1 Air Discharge Research and Development 1
1.1.1 Air Discharge Tests 2
1.1.2 Classical Discharge Theories 5
1.1.3 Physical Models of Air Discharge 7
1.1.4 Inspirations from Existing Research 8
1.2 Research Assumption of Air Insulation Prediction 10
1.2.1 Research Ideas 10
1.2.2 Implementation Method 13
1.2.3 Key Technologies 14
1.3 Contents of This Book 15
References 15

2 Theoretical Foundation of Air Insulation Prediction 19
2.1 Influence Factors of Air Discharge 19
2.1.1 Gap Structure 20
2.1.2 Applied Voltage Waveform 21
2.1.3 Atmospheric Environment 21
2.2 Energy Storage Features of Air Gap 22
2.2.1 Electric Field Features 22
2.2.2 Impulse Voltage Waveform Features 27
2.2.3 Energy Storage Features 32
2.3 Space Mapping Idea and Its Application 32
2.3.1 Basic Idea of Space Mapping 33
2.3.2 Application of Space Mapping in Insulation Prediction . . . 36
2.4 Brief Summary 40
References 41

3 Air Gap Discharge Voltage Prediction Model 43
3.1 Algorithm Selection of Prediction Model 43
3.1.1 Applications of Artificial Intelligence Algorithms 43
3.1.2 Basis for Algorithm Selection 44
3.2 Fundamental Theory of SVM 45
3.2.1 Statistical Learning Theory 46
3.2.2 Support Vector Classifier 49
3.3 Parameter Optimization Methods 54
3.3.1 Cross Validation 54
3.3.2 Grid Search Algorithm 55
3.3.3 Genetic Algorithm 55
3.3.4 Particle Swarm Optimization Algorithm 56
3.4 Feature Dimension Reduction Methods 57
3.4.1 Normalization Processing 58
3.4.2 Correlation Analysis Method 59
3.4.3 Principal Component Analysis Method 59
3.5 Sample Selection Method 60
3.6 Error Analysis Method 61
3.7 Implementation Process of the Prediction Model 61
3.8 Brief Summary 64
References 65

4 Corona Onset Voltage Prediction of Electrode Structures 67
4.1 Corona Discharge 67
4.1.1 Basic Characteristics of Corona Discharge 67
4.1.2 Corona Onset Voltage and Inception Field Strength 69
4.2 Corona Onset Voltage Prediction of Rod-Plane Electrodes 70
4.2.1 Training and Test Sample Set 70
4.2.2 SVM Prediction Results and Analysis 73
4.2.3 Comparison with Other Prediction Methods 73
4.3 Corona Onset Voltage Prediction of Stranded Conductors 88
4.3.1 Electric Field Analysis of the Stranded Conductor 89
4.3.2 Corona Onset Voltage Prediction of Single Stranded Conductors 91
4.3.3 Comparison with Other Prediction Methods 93
4.4 DC Corona Onset Voltage Prediction of Valve Hall Fittings 101
4.4.1 Corona Tests 101
4.4.2 Corona Onset Voltage Prediction 102
4.4.3 Result Analysis and Discussions 103
4.5 Brief Summary 106
References 106

5 Power Frequency Breakdown Voltage Prediction of Air Gaps 109
5.1 Air Gap Breakdown Characteristics Under Steady-State Voltage 109
5.1.1 Breakdown in Uniform Electric Field 110
5.1.2 Breakdown in Slightly Uneven Electric Field 111
5.1.3 Breakdown in Extremely Nonuniform Electric Field 112
5.2 Breakdown Voltage Prediction of Typical Short Air Gaps 113
5.2.1 Power Frequency Breakdown Voltages of Typical Air Gaps 113
5.2.2 Analysis of the Electric Field Distributions 115
5.2.3 Orthogonal Design of Training Samples 117
5.2.4 Prediction Results and Analysis 118
5.3 Breakdown Voltage Prediction of Atypical Short Air Gaps 124
5.3.1 Power Frequency Breakdown Voltages of Atypical Air Gaps 124
5.3.2 Analysis of the Electric Field Distributions 126
5.3.3 Training Sample Set 128
5.3.4 Prediction Results and Analysis 130
5.4 Brief Summary 130
References 133

6 Impulse Discharge Voltage Prediction of Air Gaps 135
6.1 Air Gap Breakdown Characteristics Under Impulse Voltages 135
6.1.1 Lightning Impulse Breakdown Characteristics 135
6.1.2 Switching Impulse Breakdown Characteristics 138
6.2 Switching Impulse Discharge Voltage Prediction 143
6.2.1 Switching Impulse Discharge Voltage Prediction of Rod-Plane and Rod-Rod Gaps 143
6.2.2 Hybrid Prediction of Switching Impulse Discharge Voltages of Different Gap Structures 147
6.3 Lightning Impulse Discharge Voltage Prediction 154
6.3.1 Positive Lightning Impulse Discharge Voltage Prediction 155
6.3.2 Negative Lightning Impulse Discharge Voltage Prediction 161
6.3.3 Prediction of Lightning Impulse Volt-Time Characteristics 164
6.4 Brief Summary 170
References 171

7 Engineering Applications of Air Insulation Prediction Model 173
7.1 Discharge Voltage Prediction of Parallel Gaps for Transmission Line Insulator String 173
7.1.1 Experimental Data and Samples 173
7.1.2 Electric Field Calculation 174
7.1.3 Prediction Results and Analysis 176
7.2 Discharge Voltage Prediction of Transmission Line-Tower Gaps 178
7.2.1 Transmission Line-Tower Air Gaps 178
7.2.2 Electric Field Calculation and Feature Extraction 182
7.2.3 Discharge Voltage Prediction 186
7.3 Discharge Voltage Prediction of Complex Gaps for Helicopter Live-Line Work 190
7.3.1 Complex Gap Discharge Voltage Prediction Method 191
7.3.2 Complex Gaps for Helicopter Live-Line Work 192
7.3.3 Electric Field Calculation and Feature Extraction 195
7.3.4 Discharge Voltage Prediction 198
7.4 Brief Summary 202
References 203

About the Authors

Dr. Zhibin Qiu was born in Jiangxi Province, China in 1991. He received the B.S. degree in electrical engineering and automation, and the Ph.D. degree in high voltage and insulation technology from the School of Electrical Engineering, Wuhan University, Wuhan, China in 2011 and 2016, respectively, where he was engaged in postdoctoral research from 2016 to 2018. He is currently a Distinguished Professor with the Department of Energy and Electrical Engineering, Nanchang University, Nanchang, China. His research interests include gas discharge, numerical analysis of electromagnetic field and its engineering applications, and the external insulation of power transmission and transformation equipment.

Prof. Jiangjun Ruan was born in Zhejiang Province, China in 1968. He received the B.S. and Ph.D. degrees in electric machine engineering from Huazhong University of Science & Technology, Wuhan, China in 1990 and 1995, respectively, and finished his postdoctoral research from Wuhan University of Hydraulic & Electric Engineering, Wuhan, China in 1998. He is currently a Professor with the School of Electrical Engineering and Automation, Wuhan University, Wuhan, China. His research interests include numerical analysis of electromagnetic field and its engineering applications, high voltage and insulation technology.

Dr. Shengwen Shu was born in Jiangxi Province, China in 1987. He received the B.S. degree in electrical engineering and automation, and the Ph.D. degree in high voltage and insulation technology from the School of Electrical Engineering, Wuhan University, Wuhan, China in 2009, and 2014, respectively. From 2014 to 2018, he is a Senior Engineer with the Electric Power Research Institute of State Grid Fujian Electric Power Co., Ltd, Fuzhou, China. Currently, he is with the College of Electrical Engineering and Automation, Fuzhou University, Fuzhou, China. His research interests include gas discharge, high voltage apparatus, and disaster prevention and reduction of power grid.

Chapter 1
Background of Air Insulation Prediction Research

1.1 Air Discharge Research and Development

Air is the most commonly used insulating medium in power systems. The issue to be solved for air insulation is how to determine the breakdown voltages of air gaps under different applied voltage types and waveforms, thus to select suitable electrode structure and gap distance for the insulation design of electrical equipment. Since air discharge theory is still not yet perfect, it is actually difficult to calculate air gap breakdown voltages accurately by theoretical methods. During the design and construction of the extra high voltage (EHV) and ultra-high voltage (UHV) transmission and transformation projects, the dielectric strengths of air gaps were mainly determined by costly and time-consuming discharge tests, or evaluated by some engineering empirical models. Therefore, how to calculate the discharge voltages of air gaps by simulation methods has become more and more attractive by researchers in the area of high voltage engineering.

In recent years, some scholars propose the concept of "computational high voltage engineering (CHVE)", aiming at promoting the transition and development of high voltage and insulation technology from a traditional subject mainly based on experimental studies to theoretical analysis and simulating calculations. The core elements of the CHVE are multi-physics field calculations and insulation discharge calculations. Along with the rapid development of numerical computation methods and the continuous improvement of computer calculation ability, the multi-physics field coupling calculation has already become the research hotspot in the area of high voltage engineering, and it plays an increasingly important role in structure design and condition assessment of electrical equipment.

Currently, various commercial software has been applied to achieve the multi-physics field simulations of high voltage equipment, including the electromagnetic field, thermal field, stress field and fluid field, etc. The simulation results can be used to check whether the insulation structure can satisfy the requirements of mechanical property and thermal property. However, even though the electric field distribution can be calculated accurately, using numerical techniques like the finite element

Z. Qiu et al., *Air Insulation Prediction Theory and Applications*, Power Systems,
https://doi.org/10.1007/978-981-10-5163-0_1

method (FEM), the withstand voltage of air insulation cannot be calculated or predicted according to the electric field simulation results. Therefore, the multi-physics field simulation without insulation calculation is incomplete, and it has long been the weak point that restricts the virtual design and intelligent manufacturing of electrical equipment.

The discharge characteristics of different dielectrics are the foundation of high voltage engineering, and their discharge voltages are important bases for insulation design of high voltage equipment. Among the commonly used dielectrics in power systems such as air, sulfur hexafluoride, transformer oil and various solid insulating materials, the understandings about air discharge are the most clearly, including the physical mechanisms, statistical law and influencing factors. Hence, air insulation calculation or prediction should be an important part and also a breakthrough point to realize the goal of CHVE.

Air discharge mechanism has been studied for more than one hundred years. But up to now, there are no perfect discharge theories and mathematical-physical models. The secret of air discharge has not yet been fully revealed from the macroscopic development process or the microscopic physical mechanism. On account of this, it has important scientific significance and application value to explore new approaches to achieve air insulation calculation or prediction, so as to provide effective supplements for the existing theories and models. The intention of this book is to propose an innovative air insulation prediction theory and introduce its implementation method and applications. We hope these studies may be helpful to lay the foundation for scientific design of air insulation structures in high voltage engineering.

Before entering the heart of this book, the existing researches about air discharge are summarized briefly. By a review of air discharge studies and developments, we will extract the related conclusions and problems, and find some useful enlightenments for the formation of the proposed air insulation prediction theory.

1.1.1 Air Discharge Tests

Air discharge results from gas ionization under the effect of an electric field. When the electric field strength applied on air dielectric exceeds a certain limit, or in other words, when the electric field energy stored in the air gap exceeds the threshold, air discharge will happen in the gap and it will lost the insulation ability. The factors influencing the electric field distribution and the atmospheric condition will change the strength of air insulation. These factors can be concluded into three aspects [1, 2]: (1) the spatial distribution of the electric field which determined by the gap configuration; (2) the temporal variation of the electric field which determined by the applied voltage shape; and (3) the air condition which defined by some atmospheric parameters. Over the years, the experimental investigations of air discharge are mainly about the influence laws of the above three sets of factors on the discharge characteristics.

In 19th century, F. Paschen summarized the relationship between the gas breakdown voltage U_b and the product of gas pressure p and gap distance d

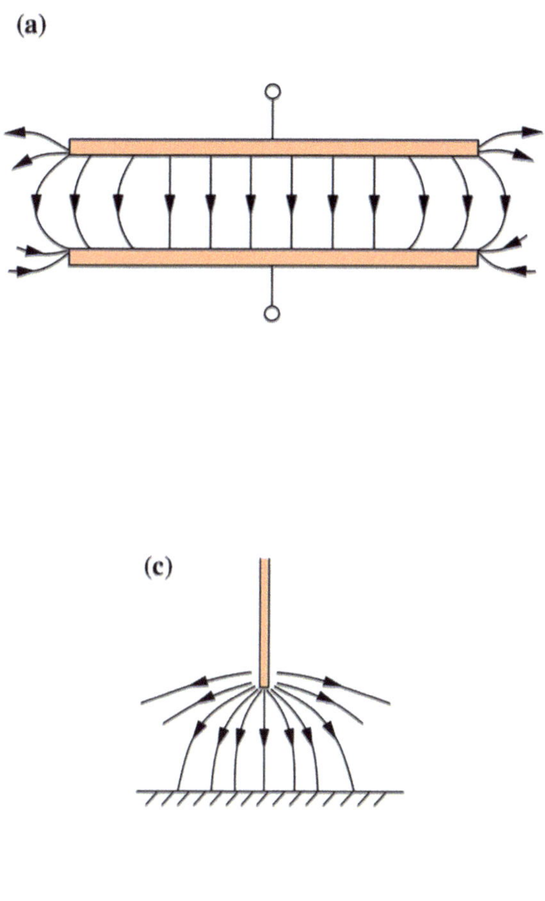

Fig. 1.1 Schematic diagrams of typical air gaps with uniform and nonuniform electric fields: **a** parallel-plate gap, **b** sphere gap, **c** rod-plane gap, **d** rod-rod gap

in uniform electric field, which is called Paschen's law. For air gap breakdown voltage in nonuniform electric field, the effect of the electric field inhomogeneity should also be considered except for the gap distance. As shown in Fig. 1.1, the middle part of the parallel-plate gap is a uniform electric field, the electric field in sphere gaps with $d/D \leq 0.5$ is a slightly uneven electric field, while D is the sphere diameter. The rod-plane gap with asymmetric structure and the rod-rod gap with symmetrical structure are two typical extremely nonuniform electric fields.

The breakdown voltage of an air gap in uniform electric field can be calculated by Paschen's law, while for that in slightly uneven electric field, the breakdown voltage can be estimated by the electric field nonuniform coefficient or the empirical formulas of streamer inception field strength. For air gaps in extremely nonuniform electric field with some specific structures, there breakdown voltages can also be estimated by several empirical formulas. In general, under the same gap distance, the air gap with more nonuniform electric field has lower breakdown voltage. For long air gaps in extremely nonuniform electric field, the average breakdown field strength under power frequency AC voltage and switching impulse voltage will decrease gradually with the increase of the gap distance. The U_b-d curve has obvious saturation phenomenon.

In order to develop EHV and UHV transmission technology, since the mid-20th century, several countries, such as the former Soviet Union, USA, Canada, Japan, South Korea, Italy, France, and China, etc., have successively constructed large high voltage test base. The researchers in these countries have carried out a large number of long air gap discharge tests, and studied the discharge characteristics of different gap configurations under various voltage types. The test data contributes to fitting numerous empirical formulas for discharge voltage calculation.

The external insulation of EHV and UHV transmission projects should be designed according to long air gap discharge characteristics under switching surges. Under normal conditions, the positive discharge voltage is lower than negative one, therefore the long air gap discharge characteristics under positive impulses have been mainly considered in insulation coordination. In summary, the main conclusions drawn from the positive switching impulse discharge tests of long air gaps can be concluded as follows.

1. Under a constant gap distance d, the relationship between the 50% discharge voltage U_{50} and the front time of the impulse voltage T_f is a U-shaped curve. The front time corresponding to the minimum U_{50} is called the critical wavefront time T_{cr}. Under a constant gap arrangement, T_{cr} has an approximate linear relation with the gap distance d. Under the impulse voltage with T_{cr}, if the curvature radius of the high voltage electrode, denoted as R, is less than a critical value R_{cr}, the U_{50} of the gap will approximately remain unchanged. For $R > R_{cr}$, U_{50} will increase obviously with the increase of R. R_{cr} is called the critical corona radius. In most practical conditions, the high voltage electrode radius is less than R_{cr} [1], while under the impulse voltage with T_{cr}, the air gap has the minimum 50% discharge voltage, which is denoted as $U_{50,\,crit+}$.
2. The gap configurations used for long air gap discharge tests mainly contain the typical gaps, like rod-plane and rod-rod gap, and engineering gaps used in actual transmission and transformation projects [3]. On the basis of experimental data, many scholars fitted the relational expression between $U_{50,\,crit+}$ (kV) and d (m). The most representative are EDF formula [4], Cortina's formula [5] and CRIEPI formula [6], etc. These formulas have different scopes of application. For the 50% discharge voltages of rod-plane gaps, denoted as U_{50RP}, under the positive standard switching impulse voltage with the waveshape of +250/2500 μs, the International Electrotechnical Commission (IEC) standard [7] considers that $U_{50RP} = 500d^{0.6}$ (kV crest, m) provides a better approximation.
3. For discharge characteristics of engineering gaps, the concept of "gap factor" [8, 9] is used to estimate the discharge voltage. The gap factor k is the ratio between the discharge voltage of an air gap to the positive rod-plane gap discharge voltage,

with identical gap spacings and submitted to the same switching impulse. k is a coefficient independent of the gap length and it relates to the gap configuration and the electrode structure [1]. IEC standard [7] recommends $U_{50} = k \cdot 500d^{0.6}$ to fit the relationship between the positive standard switching impulse discharge voltage of an engineering gap and the gap distances. Since the electrode structures of transmission tower gaps are much different from the rod-plane gap, in order to improve the fitting precision, the power exponent is usually taken as the fitting parameter as well in practical applications. That is, $U_{50} = A \cdot d^{\mathrm{B}}$ is used to fit the discharge test data, while A and B are two fitting parameters [10].

4. Air gap discharge voltage is affected by atmospheric parameters like air pressure, temperature and humidity. These three parameters are relevant to altitude. The discharge voltages under different atmospheric conditions or different altitudes can be corrected to the standard atmospheric condition using some correction formulas [11]. In addition, complex and severe ambient conditions like fog, rain, sand dust, haze, forest fire, etc. also have obvious effects on air gap discharge characteristics.

Currently, the empirical formulas fitted by discharge test data are only applicable in specific experimental conditions. These formulas only consider the influence of gap distance and do not consider the three-dimensional spatial structure in the air gap. Therefore, these fitting formulas will result in large errors when extrapolated to other conditions. Since the gap configurations in practical transmission and transformation projects are different from each other, the estimation of their discharge voltages by the gap factor also causes certain errors. Furthermore, the gap factor for each new gap configuration should be determined by a large number of discharge tests.

Among the influencing factors on air insulation strength, the air clearance and the electrode structure have decisive influences, and the shape of the impulse voltage has a significant influence [1]. Under a specific voltage waveshape and the standard atmospheric condition, it can be considered that air gap discharge voltage is determined by gap structure. The gap structure has a one-to-one corresponding relation with the static electric field distribution. Therefore, if we can establish the relationship between the static electric field distribution and the discharge voltage, it will be conductive to realize the prediction of air gap insulation strength, so as to break through the research status that gap discharge voltage should be obtained by experiments or estimated by engineering empirical models.

1.1.2 Classical Discharge Theories

Air discharge theories have been studied for more than a century. Based on large quantities of discharge experiments, J. S. Townsend studied the law of air discharge systematically in the early 20th century. He proposed the famous Townsend discharge theory [12], which interprets the gas discharge process quantitatively and lays the foundation of gas discharge theory.

One of the important successes of Townsend discharge theory is the derivation and explanation of Paschen's law. It provides a good explanation of gas discharge process of short gap with low gas pressure and uniform electric field. Townsend theory considers that the electron avalanche and the secondary electron emission are the principal mechanisms of gaseous gap breakdown. The former is caused by electron collision ionization and the latter is due to positive ions hitting the cathode. In the premise of several assumptions, Townsend theory gives the calculation formulas of discharge current and breakdown voltage. However, it is difficult to explain some discharge phenomena, such as the breakdown time-lag and the discharge channel branches. Furthermore, Townsend theory does not consider the effect of space charges caused by electron avalanches, which are very important for discharge development.

In the late 1930s, H. Raether, L. B. Loeb and J. M. Meek proposed the streamer discharge theory [13–15], which is a good supplement of Townsend theory. Streamer theory considers that air discharge is developed in the form of streamer, which is a conductive plasma channel. The physical process of gas discharge described by the streamer theory includes the development of a single electron avalanche, the transition of electron avalanche to streamer, and the streamer propagation. The streamer theory thinks that the electron collision ionization and the space photoionization are the major factors lead to self-sustained discharge, and it also takes into account the electric field distortion effect caused by electron avalanche and streamer.

When the gap distance is long, if the applied voltage is not enough to make the streamer run through the entire gap, it may still develop the breakdown process. When the streamer grows long enough, many electrons will move along the channel to the electrode. The electrons passed through the root of the channel are the most. As a result, the temperature of the streamer root rises and it will cause thermal ionization. This channel with thermal ionization process is called leader.

In 1970s, Les Renardières Group, an international group composed of researchers from several countries, established advanced discharge observation system, and studied the long air gap discharge process systematically by experimental observations [16–19]. The physical parameters including voltage, current, charge, optical spectrum and discharge pictures during the discharge process of typical long air gaps, like rod-plane gaps, were recorded. These studies basically clarify the development process of long air gap discharges and explain some discharge mechanism.

Les Renardières Group considered that the positive long air gap discharge process can be summarized as four stages, including the first corona inception, the transition from streamer to leader, the continuous leader propagation and the final jump. The negative discharge of long air gaps develops discontinuously in the way of stepped leader. Les Renardières Group summarized the formation process of one stepped leader as four stages, including the first corona inception, the pilot systems, the space leaders and the junction process. These experimental investigations preliminarily determine the basic physical parameters involved in long air gap discharge process, and reveal the physical mechanisms of different phases during the discharge.

In summary, the fundamental physical processes that are responsible for air discharge are collision ionization, photoionization, thermal ionization, cathode surface ionization, and electron attachment, recombination and diffusion, etc. Currently,

Townsend theory and streamer theory are relatively mature and widely accepted gas discharge theory. But it should be noted that the streamer theory is still not perfect, while the streamer-leader discharge process proposed by Les Renardières Group is actually a hypothesis to describe the external characteristics of long air gap discharges. Detailed introductions of air discharge process and discharge theories have been given in many books, such as [2, 12–15, 20–23], therefore they are not reviewed in detail in this book.

1.1.3 Physical Models of Air Discharge

In order to explain the experimental results of air discharge and the discharge process, many scholars carry out researches on numerical modelling and simulation of air gap discharge process. Numerous physical models have been proposed for air discharge simulations. These models are usually based on many simplifications and assumptions. The ultimate goal of numerical modelling is to simulate the behavior of long air gap discharge and predict the relevant physical parameters, especially the discharge voltage. These studies are helpful to establish an analytical measure for insulation coordination, combining simulations and experiments.

The basic idea of physical models is to establish the mathematical models or criteria of different discharge phases, and calculate the key physical parameters of each phase, so as to simulate and analyze the whole discharge process, and obtain the discharge characteristics of air gaps with different configurations, submitted to various voltage shapes and atmospheric conditions [3]. Most of the existing studies on air discharge physics were aimed at the positive discharge. Since positive air gap discharge contains different phases, the existing models include the simulation models or criteria about a single discharge phase and those considering the main phases of the discharge, which are known as self-consistent models. The single discharge phase models mainly include the corona inception criteria and models, the streamer propagation models, the leader inception criteria and models, and the leader propagation models. Some representative models for positive air discharge simulation are summarized in Table 1.1.

In Table 1.1, some models are limited to a particular discharge phase, while the others take into account the whole discharge process, but they are different in simulation of different phases. These physical models of air discharge were introduced and commented in [1, 3, 21, 54]. Most of these models are based on classical discharge theories and have rich physical connotations. They are helpful for better understanding of the physical processes involved in air discharge. Based on some simplifications and assumptions, they are applicable to calculate some physical parameters during the discharge process and predict the discharge voltages of some specific gap configurations. However, since air discharge theories are not yet perfect, some phenomena are lack of reasonable physical explanations, and some of the key physical parameters in physical models have not been unified. Most of these physical models have limitations and they are valid only under certain conditions. How to establish a scientific physical model to predict air discharge behavior still relies further studies.

Table 1.1 Physical models for positive discharge in air gaps

Discharge phase	Representative physical models
Corona inception	Critical field strength criteria (Peek's formula [24], Ortéga's formula [25], Lowke's formula [26], etc.), Critical charge criteria (Raether criterion [13], etc.), Photoionization model [27–29], Critical volume model [16], Hepworth's corona cloud model [30]
Streamer propagation	Gallimberti's equivalent electron avalanche model [31, 32], Fofana's equivalent electrical network model [33], Goelian-Lalande's model [34], Becerra-Cooray's model [35, 36], Ortéga's model [37], Arevalo's model [38]
Leader inception	Carrara's critical corona radius criterion [39], Rizk's criterion [40, 41], Local thermodynamic equilibrium (LTE) model (1500 K critical thermal ionization temperature criterion, 1 μC critical charge criterion) [32, 34]
Leader propagation	Rizk's arc model [40], LTE model [32], non-LTE model [32, 42], Fofana's equivalent electrical network model [33, 43, 44], Becerra-Cooray's model [35, 36]
Whole discharge process	Hutzler's model [45], Ortéga's model [46], Lemke's model [47], Aleksandrov's model [48], Jones's model [49], Carrara-Thione's model [39], Bazelyan's model [50], Rizk's model [40, 41], Bondiou-Gallimberti's model [34, 51], Arevalo's model [52, 53], Beroual-Fofana's model [54]

1.1.4 Inspirations from Existing Research

For air discharge researches including experiments, theories and models, the goal is to obtain air gap discharge voltage so as to guide insulation structure design of transmission and transformation projects. Some beneficial enlightenments can be derived from the pioneering investigations in the past few decades and the recent research progress, which are concluded as follows.

1. Empirical formulas obtained from discharge tests are actually a data fitting method. They are only applicable for some particular gap configurations under certain experimental conditions, and the extrapolation performance is poor. For the gap factor method which is adopted widely, the value of the gap factor k should be determined by full-scale discharge tests and varies with the gap arrangement. In fact, empirical formulas are simple relational expressions between the discharge voltage and the gap distance, the difference of electrode structure is only reflected by the gap factor. Besides, both of the gap distance and the electrode structure are decisive factors of the static electric field distribution. The fitting formulas like $U_{50} = k \cdot 500d^{0.6}$ are brief descriptions of the relevance between air gap discharge voltage and the static electric field distribution. For practical engineering air gaps, the configurations are quite complicated and with various structures. The values of the gap factor obtained by engineering experience are usually not applicable for new gap configurations. Therefore, the three-dimensional space

structure of the air gap cannot be simply characterized only by the gap distance *d* and the gap factor *k*. It is necessary to find a more effective method to describe the gap configuration or the electric field distribution.

2. According to engineering experience, the positive switching impulse discharge voltage of an arbitrary air gap can be estimated by the product of the gap factor *k* and the rod-plane discharge voltage, with an identical spacing and under the same switching impulse. The gap factor is practically independent of the gap distance [1]. That is to say, under a certain gap distance, air insulation strength has a reference value. For different electrode structures, their discharge voltages are products of this reference value and a different coefficient. This characteristic can be described as the "intrinsic property of air discharge", which is the reflection of the voltage withstand ability for the air under an external excitation. The synthetic effect of the gap structure and the air intrinsic property results in the discharge voltage, while the former is usually known, but the latter is a virtual quantity that cannot be described exactly. However, the discharge voltage can be measured by experiments. Then, the "intrinsic property of air discharge" can be described indirectly by the association relationship between the gap structure and the discharge voltage.
3. Physical models of air discharge are theoretical calculation methods based on causal analysis and model inference. Some advanced measuring and testing techniques are employed to measure the characteristic parameters during the discharge development process. Combining the physical explanations of the discharge phenomena and the progressive computational science, physical models simulate each discharge phase step-by-step through mathematical description and model inference of the causal relationship between discharge inception, development and gap breakdown, so as to reveal the internal mechanism of air discharge and calculate the related physical parameters and the discharge voltage. Although these models are based on realistic representation of the discharge, the physical processes are full of uncertain influencing factors, the discharge path is with strong randomness and some key physical parameters are with poor controllability and measurability. Therefore, it is difficult to form governing equations for rigorous mathematical description of the whole process for long air gap discharge. The existing physical models are mostly based on many empirical laws and simplifying assumptions, which will affect the model validity and the calculation accuracy.
4. Even though there are various physical models, there is no one model can be applied for discharge voltage calculation of practical engineering gap configurations. For example, the equivalent electrical network model proposed by Beroual and Fofana [54] takes the electrode geometry, the voltage waveshape and the atmospheric parameters as input data. The Peek's formula is used to judge corona inception according to high voltage electrode radius and gap distance, and a distributed-circuit model is applied to simulate the subsequent discharge process. This model has been successively applied in 50% discharge voltage calculation of typical air gaps like the rod-plane gap [55]. However, even though the U_{50} of rod-plane gaps have been calculated exactly, it still needs to use gap factor

method to estimate the discharge voltages of engineering gaps. Actually, due to the complexity and randomness of long air gap discharge process, the existing discharge theories or hypotheses are insufficient to construct governing equations to describe this physical process mathematically. Theoretical calculations lack of rigorous governing equations must be based on some engineering experience and simplifying assumptions, which are difficult to guide external insulation design effectively. Hence, how to establish an insulation calculation or prediction model applicable for engineering gaps still needs further investigations.

1.2 Research Assumption of Air Insulation Prediction

Air discharge is a very complicated nonlinear spatiotemporal evolution system. From a macro perspective, it results from the combined effect of external excitations and air dielectric. Therefore, an air discharge system can be viewed as a cause-and-effect problem. The external excitations are macroscopic factors including the applied voltage and the gap structure, which refer to the "cause". The system response is the gap discharge voltage, which refers to the "effect". Since air discharge process cannot be described by some explicit mathematical relations, this cause-and-effect relation has no governing equations. From a mathematical point of view, the external excitations of air discharge and the discharge voltage have a certain mapping relation. The related macroscopic influencing factors can be characterized by some physical or mathematical quantities, while the discharge voltage can be measured by tests. This mapping relation cannot be described by some mathematical equations with clear physical meanings. Hence, the discharge process, that is, the above mapping relation can be viewed as a grey box. Using experimental data to train the implicit functional relations between the influencing factors and the discharge voltage, it is possible to achieve air gap discharge voltage prediction under other conditions expect for the training samples. In view of this, this book proposes an innovative air insulation prediction theory, method and model, which is based on association analysis and data mining.

1.2.1 Research Ideas

As shown in Fig. 1.2, for an air discharge system, the external excitation depends on the applied voltage waveform and the gap geometry, which respectively determine the temporal variation and spatial distribution of the electric field. Both of them determine the energy storage status of the air gap before discharge inception. Under the interaction of external excitation and air dielectric, the discharge is initiated, along with the particle movement, arc development and other intermediate evolution process, which will produce electrical, optical, and thermal physical phenomena, finally results in gap breakdown.

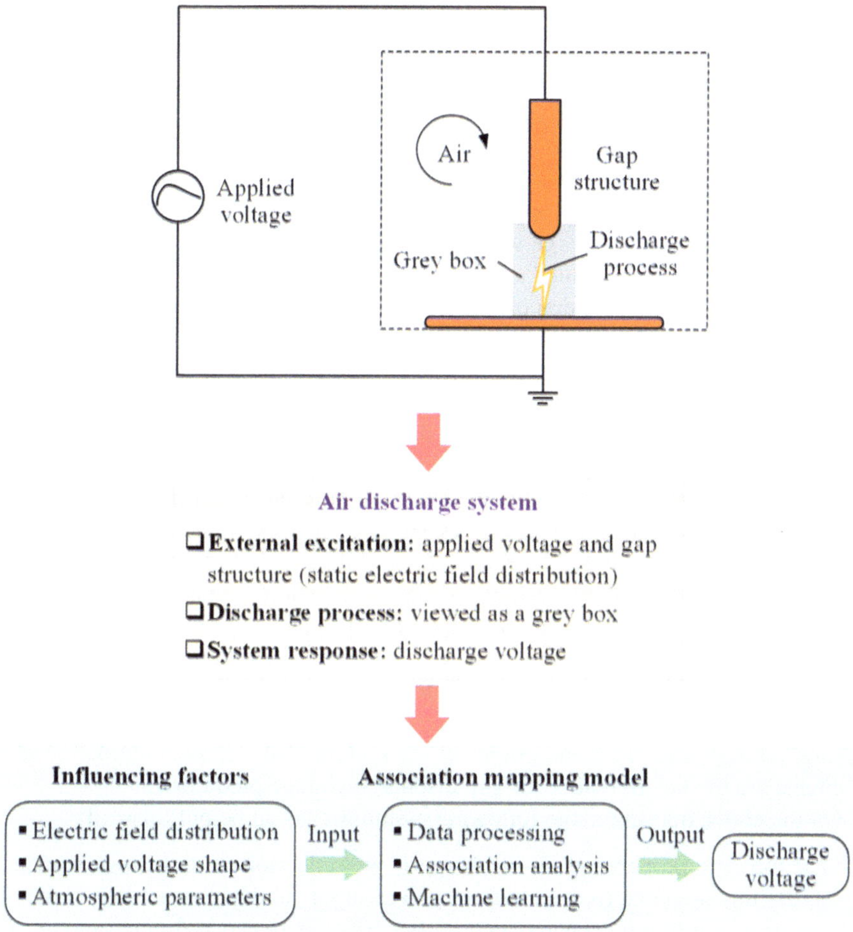

Fig. 1.2 Research assumption of air insulation prediction

The traditional physical models of air discharge are theoretical calculation methods based on causal analysis and model inference of the discharge process. The air insulation prediction theory proposed in this book is different from this, which is an intelligent prediction method based on association analysis and data mining of the relationship between the influencing factors and the discharge voltage. This research idea takes the discharge process as a grey box, and parameterizes various influencing factors by some features, which are taken as input parameters of the prediction model. Some data mining techniques, like the artificial intelligence and machine learning, are employed to establish the prediction model, and it is trained by the association relationships between the input features and output discharge voltages of the known training samples. In this way, the trained prediction model is used to forecast the dielectric strengths of other air gaps. This idea takes the deterministic macroscopic parameters, which can be quantitatively described by some physical or mathematical quantities, as the research objects. It directly establishes the relevance between the external excitations and the system response, while the former is the electric field energy storage condition before discharge inception, and the latter is the discharge voltage. With this method, we can avoid studies on the random discharge process. By macroscopic control of the applied voltage and the electric field distribution, it is able to predict air gap discharge voltage under various conditions, which is helpful to constitute the design-oriented analytical methods and theories.

For an air gap, the dispersion of its discharge voltage can be kept within a certain range under the same applied voltage waveform and the identical atmospheric environment, but the discharge process may be quite different [56]. However, the static electric field distribution before corona inception will not change [57]. It is believed that there is a strong correlation between the discharge voltage and the electric field distribution before discharge initiation. The air discharge system is translated to a mathematical problem in this book, the ideas are introduced as follows: the external excitation is characterized by the energy storage features, together with the intrinsic property of air discharge, their constitutive relations with the discharge voltage can be established to realize discharge voltage prediction. The above idea is an idealized mathematical expression, corresponding to physical mechanisms, there are three problems: (1) How to express the intrinsic property of air discharge? (2) How to establish an effective energy storage feature set to reasonably characterize the external excitation of the discharge system? (3) How to establish the constitutive relation between the input and output of the system? Hence, the solutions of these problems are the key to realize air gap discharge voltage prediction.

On the above questions, the following assumptions can be put forward:

1. The intrinsic property of air discharge is the reflection of its voltage withstand ability under an external excitation (power supply and gap structure), if the discharge results of typical gap structures (such as sphere gap, rod-plane and rod-rod gap, etc.) have been obtained, the mapping relation between the typical gap structures and their discharge voltages can be used to indirectly describe the intrinsic property of air discharge.
2. Under a constant voltage waveform, the energy storage feature set of an air gap is the characterization of its electric field distribution, and a variety of mathematical quantities can be extracted from the gap structure space to describe it as perfect as possible. For different impulse voltage waveforms, the waveform features can be used to characterize the loading process of the capacitance energy on an air gap.
3. For an air discharge system, the constitutive relation between the input and output is difficult to establish, hence this complex relation can be regarded as a grey box. The nonlinear mapping relations between the input energy storage features and the discharge voltage can be directly established by modern mathematical methods. The implicit relations between energy storage features and the discharge voltage can be formed by learning and training of the air intrinsic properties, thus

to establish the prediction theory and model of air gap insulation strength and predict the discharge voltages of complex engineering gap structures.

1.2.2 Implementation Method

In engineering, the up-and-down method is usually used in the impulse discharge test to obtain the 50% discharge voltage U_{50} of an air gap, and the implementation method is as follow. Firstly, the expected U_{50} value is estimated as U_i, and this impulse voltage is applied on the air gap. If the gap breakdown does not happen under U_i, then the next applied voltage is $U_i + \mathrm{d}U$, where $\mathrm{d}U$ is the voltage increment. Otherwise, if gap breakdown happens under U_i, the next applied voltage is $U_i - \mathrm{d}U$. The subsequent applied voltages are determined by the above method. The voltage is applied up and down until reaching enough number of times, then the voltage applied times are calculated at each voltage level to obtain the U_{50}.

Taking the idea of up-and-down test method for reference, the air gap discharge voltage prediction is converted from a regression problem to a binary classification problem. Set the discharge voltage of an air gap Ω_1 is U_b, $[(100\% - a)U_\mathrm{b}, U_\mathrm{b})$ is defined as the withstand voltage interval, denoted as -1; $[U_\mathrm{b}, (100\% + a)U_\mathrm{b}]$ is defined as the breakdown voltage interval, denoted as 1. Here, the value of a is determined by experience and the range of allowable error. By the above method, the outputs of the air insulation prediction model can be converted from the discharge voltage value to -1 and 1.

Based on the above analysis, the implementation method of the air insulation prediction proposed in this book is summarized as follows:

1. The spatial structure (spatial distribution of energy) of an air gap is characterized by the electric field features before discharge initiation, and the impulse voltage waveform features are used to characterize the accumulation process (temporal gradient) of the energy applied on the gap. The energy storage feature set composed of the above 2 types of features is used as the input of the prediction model.
2. Based on the idea of binary classification, the gap breakdown or withstanding under an applied voltage is characterized by 1 and -1 respectively, and these two values are taken as the output parameters of the prediction model.
3. The air discharge voltage prediction model is established by artificial intelligence or machine learning algorithms, so as to directly establish and the grey relation between the electric field distribution, the voltage waveform and the discharge voltage.
4. Some experimental data of typical air gaps are selected reasonably as training samples, and their energy storage features can be calculated or measured under the applied voltage respectively in the withstand interval and breakdown interval,

whose outputs correspond to −1 and 1. The prediction model is trained by the training samples, and after that it can be used to predict the discharge voltages of other gap structures.

This method assumes that the discharge voltage of an air gap under a specific voltage waveform and atmospheric environment can be kept within a certain range, and the effects of random factors in actual situations on discharge initiation position and development path are not considered. The multi-dimensional nonlinear relationship between the energy storage features of an air gap and its discharge voltage is described from mathematical perspective.

1.2.3 Key Technologies

The realization of air insulation prediction must combine theoretical analysis, mathematical modelling, numerical simulation and experimental verification. According to the above ideas, it should solve the following key technical issues.

1. Definition and extraction of electric field energy storage feature set. According to the basic laws of air discharge, the electric field distribution features and the applied voltage waveform features should be constructed, which respectively characterize the spatial structure of the air gap and the energy loading process. The calculation, extraction, selection methods of these features and the feature dimension reduction method should be studied.
2. Establishment and optimization of air insulation prediction model. Some suitable intelligent prediction algorithms should be selected to establish the prediction model. The multi-dimensional nonlinear relationships between the energy storage features of the air gap and its discharge voltage should be trained by some known experimental data. The influence rules of various factors on the prediction results should be analysed by typical prediction cases.
3. Implementation and engineering applications of air insulation prediction model. The proposed model will be used for discharge voltage prediction of air gaps with typical electrode structures, under different voltage types, thus to validate the validity of the insulation prediction theory and model. On this basis, the model should be applied to predict the discharge voltages of complex engineering gap configurations.

The research assumption of air insulation prediction proposed in this book is on the basis of gas discharge theories, computational electromagnetics, space mapping, artificial intelligence and machine learning. It involves various techniques, including electric field calculation, impulse voltage waveshape simulation, feature extraction and selection, association analysis, machine learning and intelligent prediction. The related numerical calculation and mathematical analysis methods have been applied widely in the area of electrical engineering. This book introduces these methods into the area of high voltage and insulation technology, aiming at solving the traditional problems of air insulation prediction. This work is intended to break through the research status that the insulation design relies on discharge tests. An insulation coordination and electrical design methodology combining simulations and experiments is expected to formed for fine design of external insulation structures used in transmission and transformation projects.

1.3 Contents of This Book

This book aims to provide an innovative air insulation prediction theory, method and model, and introduce its applications in discharge voltage prediction of air gaps with different configurations. Before entering the heart of this book, this chapter briefly reviews the research status of air discharge, including experiments, theories and models. The intention is not to introduce the existing air discharge theories and physical models in detail, but extract some beneficial inspirations for the research idea of this book, thus to lead to the detailed introductions of the main contents in the following chapters.

In the rest of this book, we will successively introduce the theoretical foundations of air insulation prediction, the discharge voltage prediction model, and its applications in corona onset voltage prediction of different electrode structures, steady breakdown voltage and impulse discharge voltage prediction of typical air gaps. Finally, three application cases of the proposed methodology in discharge voltage prediction of complex engineering gap configurations will be presented.

This book can provide some references for researchers and graduate students in the area of high voltage and insulation technology, gas discharge, power transmission and distribution, etc. It can also be a bibliography for engineering technical staffs related to high voltage engineering design, operation and electrical equipment manufacturing.

References

1. Thione L, Pigini A, Allen NL et al (1992) Guidelines for the evaluation of the dielectric strength of external insulation. CIGRE Brochure, Paris, France
2. Ma NX (1998) Long gap discharge. China Electric Power Press, Beijing
3. Chen WJ, Zeng R, He HX (2013) Research progress of long air gap discharge. High Volt Eng 39(6):1281–1295
4. Gallet G, Leroy G, Lacey R et al (1975) General expression for positive switching impulse strength valid up to extra long air gaps. IEEE Trans Power Appar Syst 94(6):1989–1993
5. Cortina R, Garbagnati E, Pigini A et al (1985) Switching impulse strength of phase-to-Earth UHV external insulation-research at the 1000 kV project. IEEE Trans Power Appar Syst 104(11):3161–3168
6. Kishizima I, Matsumoto K, Watanabe Y (1984) New facilities for phase-to-phase switching impulse tests and some test results. IEEE Trans Power Appar Syst 103(6):1211–1216
7. IEC 60071-2 (1996) Insulation coordination—part 2: application guide
8. Paris L (1967) Influence of air gap characteristics on line-to-ground switching surge strength. IEEE Trans Power Appar Syst 86(8):936–947

9. Paris L, Cortina R (1968) Switching and lightning impulse discharge characteristics of large air gaps and long insulator strings. IEEE Trans Power Appar Syst 87(4):947–957
10. Huo F (2012) Study on insulation characteristics and electric field distribution of long air-gaps for UHV power transmission line. Dissertation, Wuhan University
11. Sun CX, Sima WX, Shu LC (2002) Atmospheric environment and electrical external insulation. China Electric Power Press, Beijing
12. Townsend JS (1910) The theory of ionization of gases by collision. Van Nostrand Company, New York
13. Reather H (1964) Electron avalanches and breakdown in gases. Butterworth, London
14. Loeb LB, Meek JM (1941) The mechanism of the electric spark. Stanford University Press, USA
15. Meek JM, Craggs JD (1953) Electrical breakdown of gases. Oxford University Press, England
16. Les Renardières Group (1972) Research on long air gap discharges at Les Renardières. Electra 23:53–157
17. Les Renardières Group (1974) Research on long air gap discharges at Les Renardières–1973 results. Electra 35:49–156
18. Les Renardières Group (1977) Positive discharges in long air gap discharges at Les Renardières–1975 results and conclusions. Electra 53:31–153
19. Les Renardières Group (1981) Negative discharges in long air gap discharges at Les Renardières–1978 results. Electra 74:67–216
20. Raizer YP (1991) Gas discharge physics. Springer, Berlin
21. Yang JJ (1983) Gas discharge. Science Press, Beijing
22. Xu XJ, Zhu DC (1996) Gas discharge physics. Fudan University Press, Shanghai
23. Yan Z, Zhu DH (2007) High voltage and insulation technology, 2nd edn. China Electric Power Press, Beijing
24. Peek FW (1929) Dielectric phenomena in high voltage engineering. McGraw-Hill, New York
25. Ortéga P, Domens P, Dupuy J et al (1991) Long air gap discharges under non-standard positive impulse voltages. Part 2: physical interpretation. In: Paper presented at the 7th international symposium on high voltage engineering, Dresden, Germany
26. Lowke JJ, Alessandro FD (2003) Onset corona fields and electrical breakdown criteria. J Phys D Appl Phys 36(21):2673–2682
27. Nasser E, Abou-Seada M (1970) Calculation of streamer thresholds using digital techniques. IEE Conf Publ 70:534–537
28. Abdel-Salam M, Nakano M, Mizuno A (2007) Corona-induced pressures, potentials, fields and currents in electrostatic precipitator configurations. J Phys D Appl Phys 40(7):1919–1926
29. Zheng YS, He JL, Zhang B (2011) Onset criterion for positive corona in air. High Volt Eng 37(3):752–757
30. Hepworth JK, Klewe RC, Tozer BA (1972) A model of impulse breakdown in divergent field geometries. J Phys D Appl Phys 5(4):730–740
31. Gallimberti I (1972) A computer model for streamer propagation. J Phys D Appl Phys 5(12):2179–2189
32. Gallimberti I (1979) The mechanism of the long spark formation. J Phys Colloq 40(C7):193–250
33. Fofana I, Béroual A (1996) A model for long air gap discharge using an equivalent electrical network. IEEE Trans Dielectr Electr Insul 3(2):273–282
34. Goelian N, Lalande P, Bondiou-Clergerie A et al (1997) A simplified model for the simulation of positive-spark development in long air gaps. J Phys D Appl Phys 30(17):2441–2452
35. Becerra M, Cooray V (2006) A self-consistent upward leader propagation model. J Phys D Appl Phys 39(16):3708–3715
36. Becerra M, Cooray V (2006) A simplified physical model to determine the lightning upward connecting leader inception. IEEE Trans Power Deliv 21(2):897–908
37. Ortéga P, Heilbronner F, Rühling F et al (2005) Charge–voltage relationship of the first impulse corona in long airgaps. J Phys D Appl Phys 38(13):2215–2226

38. Arevalo L, Cooray V, Wu D et al (2012) A new static calculation of the streamer region for long spark gaps. J Electrostat 70(1):15–19
39. Carrara G, Thione L (1976) Switching surge strength of large air gaps: a physical approach. IEEE Trans Power Appar Syst 95(2):512–524
40. Rizk FAM (1989) A model for switching impulse leader inception and breakdown of long air gaps. IEEE Trans Power Deliv 4(1):596–606
41. Rizk FAM (1989) Switching impulse strength of air insulation: leader inception criterion. IEEE Trans Power Deliv 4(4):2187–2195
42. Gallimberti I, Bacchiega G, Bondiou-Clergerie A et al (2002) Fundamental processes in long air gap discharges. C R Physique 3(10):1335–1359
43. Fofana I, Béroual A (1995) Modelling of the leader current with an equivalent electrical network. J Phys D Appl Phys 28(2):305–313
44. Fofana I, Béroual A (1997) A predictive model of the positive discharge in long air gaps under pure and oscillating impulse shapes. J Phys D Appl Phys 30(11):1653–1667
45. Hutzler B, Hutzler-Barre D (1978) Leader propagation model for predetermination of switching surge flashover voltage of large air gaps. IEEE Trans Power Appar Syst 97(4):1087–1096
46. Ortéga P (1992) Comportement diélectrique des grands intervalles d'air soumis à des ondes de tension de polarité positive ou negative. PhD thesis, Université de Pau et des Pays de l'Adour
47. Lemke E (1967) Durchschlagmechanisms und Schlagweite-Durchschlagspannungs-Kennlinien von inhomogenen Luftfunkenstrecken bei Schaltspannungen. PhD thesis, Technical University of Dresden
48. Aleksandrov GN, Podporkyn GV (1979) Analysis of experimental data on the electric strength of long air gaps. IEEE Trans Power Appar Syst 98(2):597–605
49. Jones B (1973) Switching surges and air insulation. Phil Trans R Soc Lond A 275(1248):165–180
50. Bazelyan EM (1987) The leader of a long positive spark. Electr Tech USSR 2:47–60
51. Bondiou A, Gallimberti I (1994) Theoretical modelling of the development of the positive spark in long gaps. J Phys D Appl Phys 27(6):1252–1266
52. Arevalo L, Cooray V, Montano R (2009) Numerical simulation of long laboratory sparks generated by positive switching impulses. J Electrostat 67(2–3):228–234
53. Arevalo L, Wu D, Jacobson B (2013) A consistent approach to estimate the breakdown voltage of high voltage electrodes under positive switching impulses. J Appl Phys 114(8):083301
54. Beroual A, Fofana I (2016) Discharge in long air gaps: modelling and applications. IOP Publishing, Bristol
55. Fofana I, Beroual A, Rakotonandrasana J-H (2013) Application of dynamic models to predict switching impulse withstand voltages of long air gaps. IEEE Trans Dielectr Electr Insul 20(1):89–97
56. Wang Y, An YZ, E SL et al (2016) Statistical characteristics of breakdowns in long air gaps at negative switching impulses. IEEE Trans Dielectr Electr Insul 23(2):779–786
57. Shu SW, Ruan JJ, Huang DC et al (2015) Prediction for breakdown voltage of air gap based on electric field features and SVM. Proc CSEE 35(3):742–750

Chapter 2
Theoretical Foundation of Air Insulation Prediction

2.1 Influence Factors of Air Discharge

From a macro perspective, air discharge is a release process of the electric field energy stored in the gap, the discharge process can be considered as conversion of the capacitance energy stored in the electric field to thermal energy, luminous energy, etc., along with the transition of the discharge channel from an insulation state to a conductive plasma [1]. From a micro perspective, air discharge is a complex movement system involving a variety of particles such as electrons, ions, atoms, molecules, and photons. Under the external electric field, these particles exchange energy by colliding with each other, thus produces some microscopic physical phenomena such as ionization, recombination, photon emission and absorption [2].

Currently, classical gas discharge theories including Townsend theory, streamer theory and leader theory have been widely accepted. Under an electric field, air discharge initiates with the accelerated movement of free electrons and their collision ionization with gas molecules. Since the electron mobility is different from that of positive ions, the charge concentration will change the original electric field distribution, results in the development of ionization phenomenon along a certain channel and finally leads to gap breakdown. Based on the understanding of the above discharge process, many physical models have been proposed in the past few decades, attempting to describe the physical process of air discharge from a micro perspective, so as to explain the macroscopic laws of air discharge. However, due to the inherent complexity of the discharge process, the physical mechanisms of the air discharge are still not clear enough, and the discharge theories are not yet perfect, it is difficult to establish a generally applicable discharge model in a short term.

Actually, ionization phenomenon results from the interaction of an electric field and the air dielectric. Accurate description of air gap discharge mechanism should consider all the factors influencing the electric field distribution and the air conditions, which is obviously very difficult. However, air discharge must be caused by some

Z. Qiu et al., *Air Insulation Prediction Theory and Applications*, Power Systems,
https://doi.org/10.1007/978-981-10-5163-0_2

macroscopic factors before discharge initiation. These macroscopic factors can be classified into 3 types, including the electric field distribution, the applied voltage and the atmospheric environment. The electric field distribution can be characterized by two sets of parameters [3, 4]: (1) The static electric field distribution, determined by the shape of the electrodes, the gap length, the distance from the ground or from the surrounding electrodes, namely, the spatial scale factors; (2) The temporal variation of the electric field, namely the time scale factors, determined by the applied voltage waveform on the gap. The static electric field distribution and the applied voltage waveform are considered as 2 types of factors in this book. The static electric field distribution represents the energy storage status of an air gap in the spatial scale, the voltage waveform represents the energy loading process in the time scale, and the atmospheric environment will affect the air gap energy storage status and the energy release process after discharge initiation.

2.1.1 Gap Structure

The gap structure refers to the electrode geometry and gap distance that determines the spatial distribution of the electric field. In the study of air discharge characteristics, the related gap geometries mainly include the typical structures such as rod-plane, rod-rod, and the engineering gaps used in transmission and transformation projects. Currently, the gap structure is mainly characterized by some simple geometric parameters such as the gap length and the electrode radius. The empirical formulas used to estimate air gap discharge voltage are mostly simple relations between the discharge voltage U and the gap length d. However, the electrode shape, the surrounding grounded or energized electrodes may also affect the electric field distribution in an air gap. Hence, if the gap geometry is only characterized by the gap length, it is difficult to accurately describe the three-dimensional structure factors which affect the air gap dielectric strength. In other words, the gap length cannot accurately describe the three-dimensional electric field distribution of an air gap.

The spatial structure of an air gap can be characterized by the electric field distribution, which can be calculated by the finite element method (FEM). Besides, the electric field distribution can be characterized by some parameters such as the electric field strength, electric field energy, electric field gradient and electric field inhomogeneity, etc. which can be extracted from the FEM calculation results. Thus, infinite-dimensional electric field distribution can be transformed into finite number of electric field features. These features can be used to describe the gap structure instead of the simple geometric parameters such as the gap length and electrode radius.

2.1.2 *Applied Voltage Waveform*

Air gap breakdown voltage is related to the applied voltage type. The common voltage types in power systems contain DC voltage, power frequency AC voltage, switching impulse and lightning impulse voltage. The DC voltage and power frequency AC voltage are also collectively called steady-state voltage. They have constant waveforms with slow change rate, and therefore the discharge development time is negligible compared with it. However, the waveform and amplitude of the switching impulse and lightning impulse vary in a large range, with short duration time. Normally, the air gap dielectric strength under impulse voltages is evaluated by the discharge characteristics under 250/2500 μs standard switching impulse voltage and 1.2/50 μs standard lightning impulse voltage.

The voltage waveform determines the temporal variation of the electric field. Different applied voltages on the same air gap will lead to different breakdown voltages. Air gap breakdown characteristics under DC voltage and power frequency AC voltage only need to consider the influence of gap geometry (electric field distribution). However, for air gap discharge characteristics under switching impulse and lightning impulse voltage, the influence of voltage waveform should be taken into further consideration. The impulse voltage waveform applied on the air gap can be obtained by measurement or computer simulation, and it can be characterized by some parameters such as the voltage amplitude, voltage duration, voltage rise ratio, and voltage waveform integral, etc.

2.1.3 *Atmospheric Environment*

Atmospheric environment determines the intrinsic properties of air dielectric. It is generally believed that the atmospheric parameters influencing air gap discharge characteristics are mainly air pressure, temperature and humidity. Air pressure and temperature reflect the physical properties of air dielectric, which can be characterized by the relative air density. The humidity represents the chemical composition of air dielectric, usually considered as a separate parameter. Therefore, IEC recommended to use the relative air density and absolute humidity to characterize the influence of atmospheric conditions on air gap discharge voltage, and the standard atmospheric condition is that the air pressure $p_0 = 101.3$ kPa, the temperature $t_0 = 20$ °C, and the absolute humidity $h_0 = 11$ g/m^3.

Currently, relevant international standards have recommended the correction formulas of atmospheric parameters and altitude, which have been widely used in engineering practice. The air gap discharge voltages involved in this book have been corrected to standard atmospheric conditions, and therefore air insulation predictions in different atmospheric environment have not yet been taken into consideration.

2.2 Energy Storage Features of Air Gap

2.2.1 Electric Field Features

For an air gap, the external electric field will affect each discharge phase, including the first corona inception, the streamer propagation, the transition of streamer to leader, and the continuous leader propagation. At the same time, the space charge produced during the discharge development will also distort the interelectrode electric field distribution. There is an interactive relationship between the electric field and the discharge process, which is difficult to be described by explicit mathematical expressions. Under a specific voltage waveform, the static electric field distribution of an air gap is constant before corona inception, but the discharge development process may be quite different [5]. Therefore, we can directly establish the grey correlation between the static electric field distribution and the discharge voltage, and avoid studying the discharge process with complicated physical mechanism and large dispersion. Since the electric field is of infinitely dimensional distribution, it is necessary to extract the feature set strongly related to the discharge voltage, thus it can be used in the discharge voltage prediction model.

2.2.1.1 Definitions of Electric Field Features

Taking a hemispherical rod-plane air gap as example, the definitions of the electric field features are introduced as follows. As shown in Fig. 2.1, the hemispherical rod radius is R, and the gap distance is d. If the rod electrode is applied high voltage and the plane electrode is grounded, the electric field concentration will generate around the rod electrode surface, and the maximum field strength will appear at the rod tip, besides, the field strength will decrease gradually along the shortest path from the rod to the plane electrode. In ideal conditions, the rod-plane air gap discharge will develop along the axis of the gap. Here, a cylindrical region with a radius of R between the rod and plane electrodes is defined as the discharge channel, and the shortest geometric path along the axis of the gap is assumed to be the shortest discharge path. It is known that the discharge development process mainly depends on the electric field distributions in the discharge channel and along the shortest discharge path. The electric field features in this region can be used as the input parameters of the air gap discharge voltage prediction model, since they have strong correlations with the air gap breakdown voltage. In addition, the corona onset voltage is related to the electrode surface states, and therefore some features is also defined on the electrode surface for corona onset voltage prediction.

The electric field distribution of an air gap can be obtained by FEM calculation, after post-processing of the calculation results, the features used to characterize the electric field distribution can be extracted. According to the discharge channel and the shortest discharge path shown in Fig. 2.1, 4 kinds of electric field features are defined as follows:

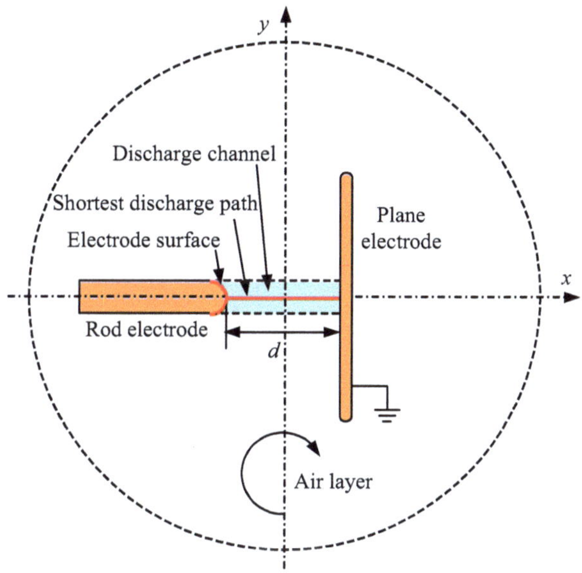

Fig. 2.1 Schematic diagram of the discharge channel, the shortest discharge path and the electrode surface of the rod-plane air gap

1. Electric field strength, including the maximum electric field strength E_{m} and the average electric field strength E_{a} in the discharge channel.

$$E_{\mathrm{m}} = \max\ E_i \quad (i = 1, 2, \ldots, n) \tag{2.1}$$

$$E_{\mathrm{a}} = \sum_{i=1}^{n} E_i / n \tag{2.2}$$

where E_i is the electric field strength of the ith volume/area element, n is the amount of the volume/area elements.

2. Electric field energy, including the total electric field energy W and energy density W_{d} in the discharge channel.

$$W = \sum_{i=1}^{n} W_i = \sum_{i=1}^{n} \left(\frac{1}{2} \varepsilon_0 E_i^2 V_i \right) \tag{2.3}$$

$$W_{\mathrm{d}} = W / \sum_{i=1}^{n} V_i \tag{2.4}$$

where ε_0 is the permittivity of vacuum, W_i and V_i are the energy and volume of the ith volume element, respectively.

3. Electric field gradient on the shortest discharge path, including the maximum value $E'_{\max}$, the minimum value $E'_{\min}$, and the average value E'_{ave}.

$$E'_{\max} = \max(|-\mathrm{grad}E_i|) \quad (i = 1, 2, \ldots, n) \tag{2.5}$$

$$E'_{\min} = \min(|-\mathrm{grad}E_i|) \quad (i = 1, 2, \ldots, n) \tag{2.6}$$

$$E'_{\mathrm{ave}} = \sum_{i=1}^{n} (|-\mathrm{grad}E_i|)/n \tag{2.7}$$

4. Electric field inhomogeneity, characterized by some scaling parameters related to the electric field strength, electric field energy and electric field gradient in the discharge channel or on the shortest discharge path. These features can be divided into relative scaling parameters and absolute scaling parameters. Relative scaling parameters include the distortion factor of the electric field E_{d}, the volume ratio $V_{\mathrm{r}x}$ and energy ratio $W_{\mathrm{r}x}$ of the region in the discharge channel whose electric field strength exceeds $x\%E_{\mathrm{m}}$, the length ratio $E'_{\mathrm{r}x}$ of the path whose electric field gradient exceeds $x\%E'_{\max}$ on the shortest discharge path.

$$E_{\mathrm{d}} = (E_{\mathrm{m}} - E_{\mathrm{a}})/E_{\mathrm{a}} \tag{2.8}$$

$$V_{\mathrm{r}x} = \sum_{xi=1}^{xn} V_{xi} / \sum_{i=1}^{n} V_i \tag{2.9}$$

$$W_{\mathrm{r}x} = \sum_{xi=1}^{xn} W_{xi} / W \tag{2.10}$$

$$E'_{\mathrm{r}x} = L_x / L \tag{2.11}$$

where V_{xi} and W_{xi} are respectively the volume and energy of the ith volume element whose electric field strength exceeds $x\%E_{\mathrm{m}}$ and xn is the amount of those volume elements, L_x is the length of the path whose electric field gradient exceeds $x\%E'_{\max}$ and L is the total length of the shortest path, that is, the gap distance d. In this book, $x\%$ includes 90, 75, 50 and 25%.

Absolute scaling parameters include the volume ratio and the energy ratio of the region whose electric field strength exceeds 7 and 24 kV/cm, namely, $V_{\mathrm{r}24}$, $V_{\mathrm{r}7}$ and $W_{\mathrm{r}24}$, $W_{\mathrm{r}7}$, the length and length ratio of the path whose electric field strength exceeds 7 and 24 kV/cm, namely, L_{24}, L_7 and $L_{\mathrm{r}24}$, $L_{\mathrm{r}7}$, whose calculation formulas are similar to Eqs. 2.9–2.11. It should be noted that 7 kV/cm is the electric field strength for the stable propagation of a streamer [6], and 24 kV/cm is the field strength at the ionosphere boundary in standard atmospheric condition, which can be calculated by Eqs. 2.12 and 2.13 [7], where α is the ionization coefficient and η is the attachment coefficient.

$$\frac{\alpha}{\delta} = \begin{cases} 3632\exp\left(-168.0\dfrac{\delta}{E}\right), & 1.9 \leq \dfrac{E}{\delta} \leq 45.6 \\ 7358\exp\left(-200.8\dfrac{\delta}{E}\right), & 45.6 \leq \dfrac{E}{\delta} \leq 182.4 \end{cases} \tag{2.12}$$

$$\frac{\eta}{\delta} = 9.865 - 0.541\frac{E}{\delta} + 1.145 \times 10^{-2}\left(\frac{E}{\delta}\right)^2 \tag{2.13}$$

In Eqs. 2.12 and 2.13, both α and η are in cm^{-1}, E is the electric field in kV/cm, δ is the relative air density, and it can be calculated by Eq. 2.14.

$$\delta = \frac{p}{T} \cdot \frac{T_0}{p_0} = \frac{p}{p_0} \cdot \frac{273 + t_0}{273 + t} = \frac{2.89p}{273 + t} \tag{2.14}$$

where p is the air pressure in kPa, t is the temperature in °C. $p_0 = 101.3$ kPa and $t_0 = 20$ °C are air pressure and temperature in standard atmospheric condition.

The electric field features on the electrode surface are as follows:

1. The maximum electric field strength E_{ms}, whose definition is similar to Eq. 2.1, the difference is that E_i is the electric field strength of the ith area element, n is the amount of the area elements.
2. The area of the region S_x on the high voltage electrode surface whose electric field strength exceeds $x\%E_m$.

$$S_x = \sum_{j=1}^{m} S_j \tag{2.15}$$

where S_j is the area of the jth area element whose electric field strength exceeds $x\%E_m$, and m is the amount of those area elements.

3. Area ratio of the region S_{rx} on the high voltage electrode surface whose electric field strength exceeds $x\%E_m$.

$$S_{rx} = S_x / \sum_{i=1}^{n} S_i \tag{2.16}$$

where S_i is the area of the ith area element, and n is the total number of the area elements.

Similarly, for the electrode surface features, $x\%$ includes 90, 75, 50 and 25%.

The above features contain three spatial dimensions including line, area and volume. These features describe the electric field distribution in various aspects, such as the electric field strength, energy, gradient, inhomogeneity (distortion factor) and so on, which can reflect the spatial structure of the air gap.

2.2.1.2 Extraction Method of Electric Field Features

Taking rod-plane gap shown in Fig. 2.2a for example, the extraction method of the electric field features is introduced below. A two-dimensional axisymmetric simulation model of the rod-plane gap is established by a finite element analysis software, the diameter of the hemispherical rod electrode is $2R$ and the gap distance is d. Research shows that the rod-plane gap discharge will develop along the axis from the rod electrode to the plane electrode in ideal conditions [8], the discharge channel and the shortest discharge path are defined in Fig. 2.2a, while the discharge channel is the interelectrode area with the width of $2R$. The discharge channel and the shortest discharge path are refined meshed, as shown in Fig. 2.2b. Taking the rod radius $R =$ 10 mm and the gap distance $d = 3$ cm for example, when the rod electrode is applied a unit potential 1 V, the plane electrode and the truncated air boundary is applied zero potential, the electric field distribution calculation result is shown in Fig. 2.2c.

The static electric field distributions of air gaps with different electrode structures can be calculated by a similar method. The electric field calculation results are post-processed to extract some original data like the values of the electric field strength on the node, element and path, the coordinates of the nodes and elements' barycenter, and the element area or volume, etc. These initial data will be input to the MATLAB software to calculate the electric field features of each gap under the applied voltage, according to their definitions and calculation formulas.

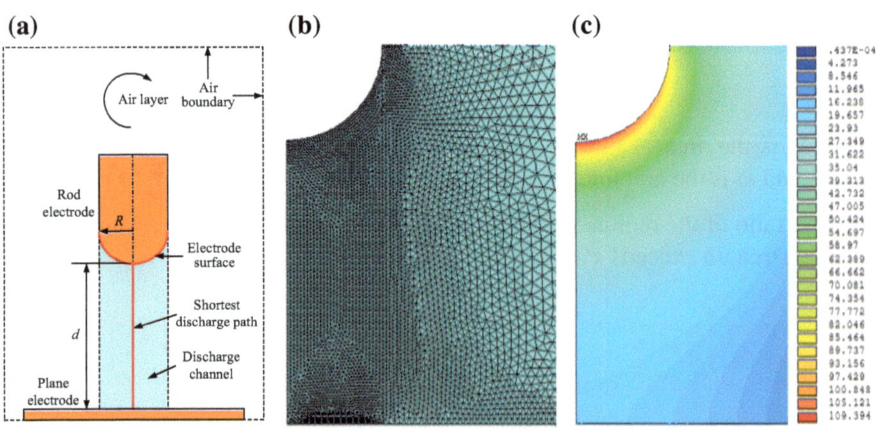

Fig. 2.2 Static electric field simulation model of the rod-plane gap and the calculation result: **a** simulation model, **b** refined mesh, **c** electric field distribution

2.2.2 *Impulse Voltage Waveform Features*

2.2.2.1 Definitions of Impulse Voltage Waveform Features

Except for the static electric field distribution, the applied voltage waveform may also affect the air gap discharge voltage obviously. Here, the applied voltage waveforms refer to the impulse voltage. Since the duration of steady-state voltages, like DC voltage and power frequency AC voltage, is much longer than the time required for gap breakdown, their effects on the discharge characteristics can be neglected. The standard switching impulse and lightning impulse voltage waveform recommended in IEC standard [9] are shown in Fig. 2.3a, b.

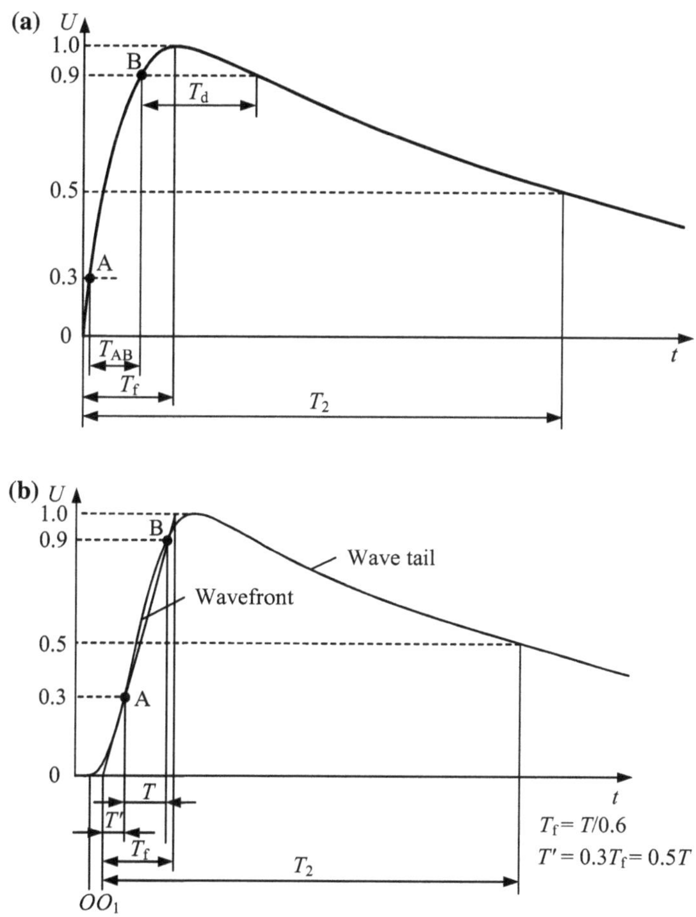

Fig. 2.3 Impulse voltage waveform: **a** standard switching impulse voltage waveform, **b** standard lightning impulse voltage waveform

The change rate, peak value and duration time of the impulse voltage will affect the air gap discharge characteristics. According to the impulse voltage waveform characteristics and its influence on air gap breakdown, the following features are used to characterize the voltage waveform:

1. The maximum value of the impulse voltage $U_{\max}$, characterizing the upper limit of the instantaneous energy applied on the air gap.

$$U_{\max} = \max(u(t)) \tag{2.17}$$

2. The time features, including the front time T_{f} and the time to half-value T_2. These two features respectively describe the rise time and duration time of the impulse voltage waveform.

For switching impulse voltage, the front time T_{f} is the time interval from the actual origin to the peak time of the voltage waveform, the time to half-value T_2 is the time interval from the actual origin to the time when the voltage firstly decreases to the half peak value.

For lightning impulse voltage, it is difficult to determine the origin of the waveform since the waveshape extracted by an oscilloscope is usually fuzzy near the zero point or there is an initial oscillation. Besides, the voltage waveform is relatively flat near the peak and therefore the peak time is also difficult to determine. Hence, IEC standard recommended the following method to determine the time features. As shown in Fig. 2.3b, a straight line is drawn to connect point A and point B, respectively corresponding to $30\%U_{\max}$ and $90\%U_{\max}$. The intersection point of this line with the time axis is set as the virtual origin O_1, then the front time T_{f} and the time to half-value T_2 can be calculated according to O_1. Set the time interval from point A to point B as T, then $T_{\mathrm{f}} = 1/0.6T$, and T_2 is the time interval from O_1 to the time when the voltage firstly decreases to the half peak value.

3. The average rise rate of the voltage waveform du/dt, which means the slope of the fitting straight line between $30\%U_{\max}$ and $90\%U_{\max}$. This feature characterizes the steepness or the rise speed of the impulse voltage waveform, which has an important effect on the generation of effective free electrons. For lightning impulse voltage waveform, the voltage rise rate is:

$$\frac{\mathrm{d}u}{\mathrm{d}t} = \frac{U_{\max}}{T_{\mathrm{f}}} \tag{2.18}$$

Since the impulse voltage at the rising stage is approximately linear with the time, for the switching impulse voltage waveform, du/dt also can be calculated by Eq. 2.18.

4. The voltage integral S, which means the integral of the impulse voltage waveform with respect to time, also can be described by the curved surface area under the voltage waveform. The voltage integral characterizes the accumulative effect of the applied energy on the gap breakdown.

$$S = \int_{0}^{+\infty} u(t)\mathrm{d}t \tag{2.19}$$

The above-mentioned 5 basic features describe the global properties of the impulse voltage waveform. Furthermore, researches have shown that the impulse discharge characteristics of an air gap almost only depend on the voltage features near a certain percent of $U_{\max}$. For example, Harada studied the effects of the switching impulse voltage waveform on the discharge characteristics by experiments [10]. The results showed that the 50% discharge voltage U_{50} depends largely on the voltage steepness near 90% $U_{\max}$. Schneider and Zaffanella assumed the region above 80% $U_{\max}$ is the equivalent wave for the double exponential wave [11]. Menemenlis and Isaksson think that the most important part of the wavefront is the region above 85% $U_{\max}$, and the shape of the front below 85% $U_{\max}$ has no effect on U_{50} of the gap [11]. They proposed the concept of "85% slope front time". It can be seen that no final conclusion has yet been reached on when the impulse voltage waveform affects the gap breakdown.

Hence, the following additional features related to $x\% U_{\max}$ are defined and applied to characterize the local properties of the impulse voltage waveform.

1. The slope at the moment of $x\% U_{\max}$, called k_x.

$$k_x = \left.\frac{\mathrm{d}u}{\mathrm{d}t}\right|_{t=t_x} \tag{2.20}$$

where t_x is the moment when the impulse voltage rises to $x\% U_{\max}$ during the wavefront.

2. The time interval T_x for which the impulse voltage exceeds $x\% U_{\max}$.

$$T_x = t'_x - t_x \tag{2.21}$$

where t'_x is the moment when the voltage decreases to $x\% U_{\max}$ during the wave tail.

3. The time interval $T_{\mathrm{f}x}$ for which the impulse voltage exceeds $x\% U_{\max}$ during the wavefront.

$$T_{\mathrm{f}x} = T_{\mathrm{f}} - t_x \tag{2.22}$$

4. The voltage integral S_x over the time interval T_x.

$$S_x = \int_{t_x}^{t'_x} u(t)\mathrm{d}t \tag{2.23}$$

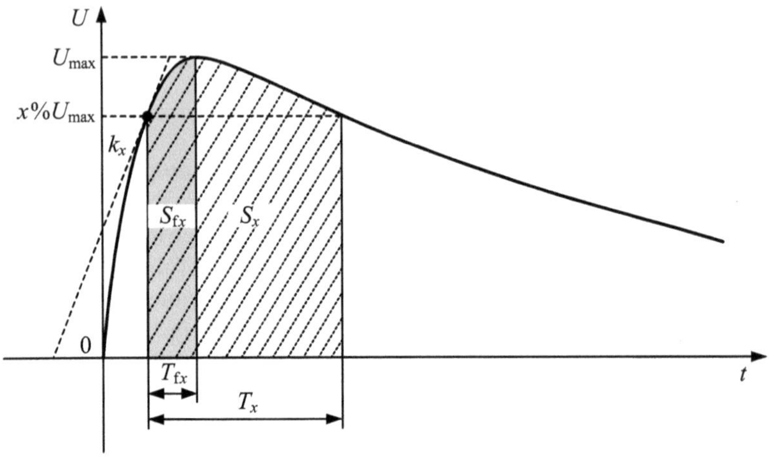

Fig. 2.4 Definitions of the additional impulse voltage waveform features

5. The voltage integral S_{fx} over the time interval T_{fx}.

$$S_{fx} = \int_{t_x}^{T_f} u(t)\mathrm{d}t \tag{2.24}$$

The definitions of the above additional features are shown in Fig. 2.4 intuitively. k_x is the feature for steepness, T_x and T_{fx} are time features, S_x and S_{fx} are voltage integral features. $x\%$ includes 90, 75 and 60%.

2.2.2.2 Calculation Method of Impulse Voltage Waveform Features

The impulse voltage waveform is usually simulated by a double exponential function. The full wave expression is

$$u(t) = A(\mathrm{e}^{-\alpha t} - \mathrm{e}^{-\beta t}) \tag{2.25}$$

where A is the amplitude coefficient, α and β are respectively the reciprocal of wave tail and wave front constant. The three parameters are usually unknown coefficients.

1. Switching impulse voltage waveform

According to the definition of switching impulse voltage waveform, there are three constraints: (1) the voltage value at the moment of T_f is the crest voltage U_{max}; (2) the voltage value at the moment of T_2 is $1/2U_{max}$; (3) the derivative of $u(t)$ at the moment of T_f is 0. The constraints can be expressed as follows:

$$
\begin{aligned}
&u(T_{\mathrm{f}}) = A(\mathrm{e}^{-\alpha T_{\mathrm{f}}} - \mathrm{e}^{-\beta T_{\mathrm{f}}}) = U_{\max} \\
&u(T_2) = A(\mathrm{e}^{-\alpha T_2} - \mathrm{e}^{-\beta T_2}) = 0.5 \times U_{\max} \\
&\left.\frac{\mathrm{d}u}{\mathrm{d}t}\right|_{t=T_{\mathrm{f}}} = A(\beta \mathrm{e}^{-\beta T_{\mathrm{f}}} - \alpha \mathrm{e}^{-\alpha T_{\mathrm{f}}}) = 0
\end{aligned} \tag{2.26}
$$

In practical application, the impulse voltage waveform is usually described by known $U_{\max}$, T_{f} and T_2. The voltage rise rate can be calculated by Eq. 2.18. By solving the nonlinear Eq. 2.26 using the least square method, the three unknown parameters A, α and β can be obtained, and therefore the impulse voltage waveform features can be calculated according to the aforementioned formulas.

After A, α and β are obtained, the voltage integral S can be further calculated by the definite integral, that is,

$$
S = \int_0^{+\infty} u(t)\mathrm{d}t = \int_0^{+\infty} A(\mathrm{e}^{-\alpha t} - \mathrm{e}^{-\beta t})\mathrm{d}t = A\left(\frac{1}{\alpha} - \frac{1}{\beta}\right) \tag{2.27}
$$

In addition, in order to calculate the additional features corresponding to $x\%U_{\max}$ moments, t_x and t'_x should be firstly calculated, which are the two roots of the following equation:

$$
u(t) = A(e^{-\alpha t} - e^{-\beta t}) = x\% \times U_{\max} \tag{2.28}
$$

After then, the additional features can be calculated according to Eqs. 2.20–2.24.

2. Lightning impulse voltage waveform

According to the definition of lightning impulse voltage waveform, set the corresponding moments of the virtual origin O_1 and the voltage peak are T_0 and T_{m}, there are five constraints for lightning impulse wave: (1) the voltage value at the moment of T_{m} is the crest voltage $U_{\max}$; (2) the voltage value at the moment of $T_0 + T_2$ is $0.5U_{\max}$; (3) the voltage value at the moment of $T_0 + 0.3T_{\mathrm{f}}$ is $0.3U_{\max}$; (4) the voltage value at the moment of $T_0 + 0.9T_{\mathrm{f}}$ is $0.9U_{\max}$; (5) the derivative of $u(t)$ at the moment of T_{m} is 0. These constraints can be expressed as follows:

$$
\begin{aligned}
&u(T_{\mathrm{m}}) = A(\mathrm{e}^{-\alpha T_{\mathrm{m}}} - \mathrm{e}^{-\beta T_{\mathrm{m}}}) = U_{\max} \\
&u(T_0 + T_2) = A(\mathrm{e}^{-\alpha(T_0+T_2)} - \mathrm{e}^{-\beta(T_0+T_2)}) = 0.5U_{\max} \\
&u(T_0 + 0.3T_{\mathrm{f}}) = A(\mathrm{e}^{-\alpha(T_0+0.3T_{\mathrm{f}})} - \mathrm{e}^{-\beta(T_0+0.3T_{\mathrm{f}})}) = 0.3U_{\max} \\
&u(T_0 + 0.9T_{\mathrm{f}}) = A(\mathrm{e}^{-\alpha(T_0+0.9T_{\mathrm{f}})} - \mathrm{e}^{-\beta(T_0+0.9T_{\mathrm{f}})}) = 0.9U_{\max} \\
&\left.\frac{\mathrm{d}u}{\mathrm{d}t}\right|_{t=T_{\mathrm{m}}} = A(\beta \mathrm{e}^{-\beta T_{\mathrm{m}}} - \alpha \mathrm{e}^{-\alpha T_{\mathrm{m}}}) = 0
\end{aligned} \tag{2.29}
$$

The equation set 2.29 contains 5 equations and 5 unknown parameters. A, α and β can be obtained by known U_{max}, T_{f} and T_2, and then the voltage rise rate du/dt, the voltage integral S and the additional features corresponding to $x\%U_{\mathrm{max}}$ moment can be calculated by the same way with the switching impulse voltage waveform.

2.2.3 Energy Storage Features

In conclusion, the energy storage features defined in this book are summarized in Table 2.1, including the electric field features and the voltage waveform features. The electric field features are defined in three spatial positions, namely, the discharge channel, the shortest discharge path and the electrode surface, involving three spatial dimensions, namely, the line, area and volume. The impulse voltage waveform features are defined in two aspects to describe the global and local properties of the waveform, and these features are divided into the basic features and the additional features. In practical applications, different types of features can be selected as required. For example, in order to analyze air gap breakdown characteristics under steady-state voltage, like DC voltage and power frequency AC voltage, the voltage waveform features are unnecessary to be taken into account because of the constant applied voltage waveform. For the corona onset voltage prediction, 3 kinds of electric field features are considered, including the discharge channel, the shortest discharge path and the electrode surface. However, for the breakdown voltage prediction, only the electric field features in the discharge channel and on the shortest discharge path are considered.

2.3 Space Mapping Idea and Its Application

The air insulation prediction method proposed in this book takes reference of the space mapping idea. The air discharge prediction problem is reconsidered from a mathematical perspective, and the complex air discharge system is transformed to a many-to-one correlation mapping problem. Air gap discharge voltage is hopeful to be realized by establishing the implicit mapping relation between the system inputs, which are the energy storage features and the air intrinsic properties, and the output discharge voltage. Meanwhile, the relationship between the fine physical model and the proposed air insulation prediction model can also be explained by the space mapping idea.

Table 2.1 Energy storage features of air gaps

Classification	Energy storage features
Electric field features	Discharge channel: maximum and average values of the electric field strength E_{m}, E_{a}; electric field energy W and energy density W_{d}; distortion factor of the electric field E_{d}; volume ratio $V_{\mathrm{r}x}$ and energy ratio $W_{\mathrm{r}x}$ of the region exceeds $x\%E_{\mathrm{m}}$; volume ratios $V_{\mathrm{r}7}$, $V_{\mathrm{r}24}$ and energy ratios $W_{\mathrm{r}7}$, $W_{\mathrm{r}24}$ of the region exceeds 7 and 24 kV/cm
	Shortest discharge path: maximum, minimum and average values of the electric field gradient, E'_{max}, E'_{min} and E'_{ave}; length ratio $E'_{\mathrm{r}x}$ of the path exceeds $x\%E'_{\mathrm{max}}$; length L_7, L_{24} and length ratios $L_{\mathrm{r}7}$, $L_{\mathrm{r}24}$ of the path exceeds 7 and 24 kV/cm
	Electrode surface: maximum electric field strength E_{ms}; superficial areas S_{90}, S_{75}, S_{50}, S_{25} and area ratios $S_{\mathrm{r}90}$, $S_{\mathrm{r}75}$, $S_{\mathrm{r}50}$, $S_{\mathrm{r}25}$ of the region exceeds $x\%E_{\mathrm{ms}}$
Voltage waveform features	Basic features: voltage amplitude U_{max}; front time T_{f} and time to half-value T_2; voltage rise rate du/dt; voltage integral S
	Additional features: wavefront slope k_x at $x\%U_{\mathrm{max}}$; time interval T_x for which the voltage exceeds $x\%U_{\mathrm{max}}$; time interval $T_{\mathrm{f}x}$ for which the voltage exceeds $x\%U_{\mathrm{max}}$ during the wavefront; voltage integral S_x over T_x; voltage integral $S_{\mathrm{f}x}$ over $T_{\mathrm{f}x}$

2.3.1 Basic Idea of Space Mapping

Space mapping techniques were firstly proposed by J. W. Bandler in 1994 [12]. The basic idea is to establish the mapping relations between the design parameters of the coarse model and the fine model by some mathematical method, and constantly optimize these mappings by iteration, so as to find the optimal solution of the fine model.

The space mapping algorithm involves two models, one is the fine model that is accurate but less efficient, the other is the coarse model that is not accurate enough but highly efficient. It is assumed that the performance of a physical object depends on a series of parameters, in order to find the optimal parameter settings, it is necessary to find the model response corresponding to some intermediate parameter set during the optimization process, which requires to solve functions by some mathematical optimization algorithms. The solving process usually costs too much, such as large amounts of calculation and low computational efficiency, so that the traditional optimization algorithms are unrealistic in the applications. Therefore, the goal of space mapping technology is to seek a shortcut to obtain the relevant information of the optimal parameters of the original model by a lower cost and less accurate coarse system with the same physical mechanism [13].

The optimization process of a microwave passive circuit is taken as an example to introduce this method as follows [14].

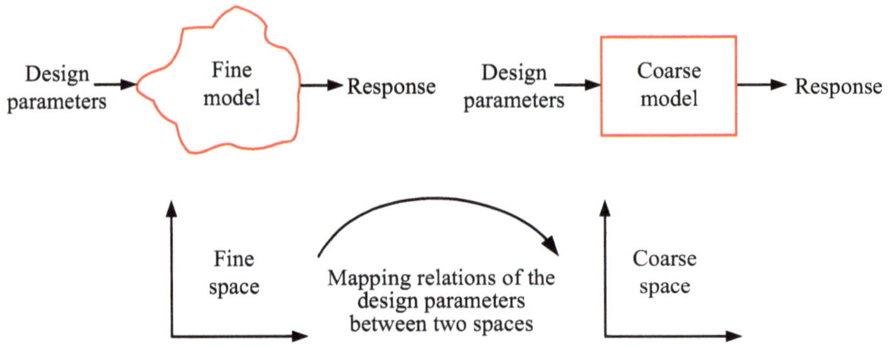

Fig. 2.5 Mapping relations of the design parameters between fine space and coarse space

The electromagnetic simulation model of a circuit can reflect its actual characteristics accurately and comprehensively, which is called the fine model. The empirical model or the equivalent circuit model can only approximately or partly reflect its actual characteristics, which is called the coarse model. As shown in Fig. 2.5, it is assumed that there are certain mapping relations between the design parameters of the fine model and the coarse model with the same response, these mapping relations can be explicit (can be expressed by mathematical methods), implicit (cannot be expressed by mathematical methods), linear or nonlinear. The space mapping method is used to optimize the circuit. Firstly, the optimal design parameters of the coarse model are obtained to satisfy the design requirements. Then, the optimal design parameters of the coarse model and the inverse mapping of the established mapping relations are used to predict the design parameters of the fine model. If the response obtained by the simulation in fine model space, using the predicted design parameters, can satisfy the design requirements, then, it is considered that the predicted design parameters of the fine model are the optimal design values of the circuit, and the mapping relations can accurately reflect the relations between these two design parameters in two spaces. Otherwise, the mapping relations between the two design parameter spaces are constantly updated and improved, and the new predicted design parameters of the fine model are predicted and verified at the same time. The optimization process is finally convergent and successful until the optimal design values of the circuit satisfy the design requirements.

The circuit optimization problem can be described by the following mathematical expression:

$$x_f^* = \arg \min_{x_f} U(R_f(x_f)) \tag{2.30}$$

Here, $R_f \in \Re^{m\times 1}$, represents a response vector with m response points in the fine space, $x_f \in \Re^{n\times 1}$, represents a vector with n design parameters of the fine model; $U(R_f(x_f))$ is a function of error, namely the objective function; x_f^* is the undetermined

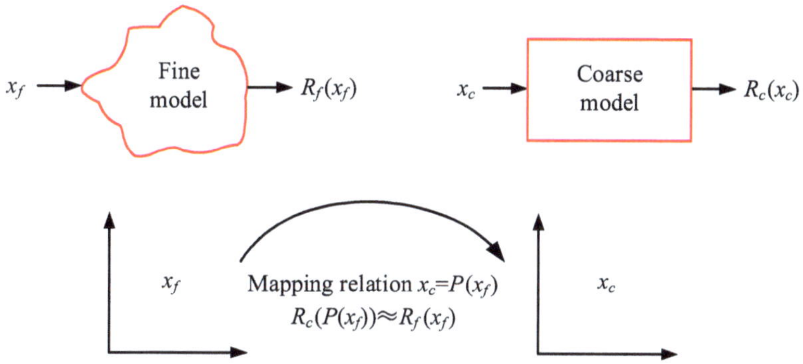

Fig. 2.6 Parameter space mapping relation

optimal design parameter of the fine model space, which is assumed to be unique. Similarly, R_c, x_c and x_c^* in coarse model space have similar definitions.

Suppose there is a space mapping P between the design parameters of the fine model and the coarse model, and it makes the responses of the two spaces match each other.

$$x_c = P(x_f) \tag{2.31}$$

$$R_c(P(x_f)) \approx R_f(x_f) \tag{2.32}$$

The mapping relation of the two spaces is shown in Fig. 2.6.

The design parameters of the fine model x_f and the coarse model x_c whose responses match each other are often referred as correlation parameters. In order to establish a mapping P, at least one pair of the correlation parameters must be known firstly, usually obtained by approaching the response of the coarse model space with that of the fine model space, so as to get the relevant design parameters of the coarse model corresponding to those of the fine model. This process is called parameter extraction (PE) and its mathematical expression is as follow:

$$x_c = \arg \min_{x_c} \left\| R_f(x_f) - R_c(x_c) \right\| \tag{2.33}$$

Parameter extraction technique is a key step in the space mapping method. The non-uniqueness of the extraction process may lead to failure of the algorithm, and various parameter extraction methods have been proposed in previous researches, including the multipoint PE, statistical PE, penalized PE, PE involving frequency mapping, gradient PE and PE by simulation software, etc. The parameter extraction avoids direct optimization of the fine model, which transfers the optimizations to the coarse model space, while the simulation of the fine model space is only used to provide the required data samples and verify whether the algorithm is convergent.

If the optimal design parameters of the coarse model x_c^* have been obtained, the design parameters of the fine model $\bar{x}_f$ can be predicted by the inverse mapping of the established mapping relation P:

$$\bar{x}_f = P^{-1}(x_c^*) \tag{2.34}$$

If the response of the fine model space $R_f(\bar{x}_f)$ meets the design specifications under the predicted fine model design parameters $\bar{x}_f$, then the predicted parameters are the final optimal design parameters of the circuit, namely $x_f^* = \bar{x}_f$, and the mapping relation P can accurately reflect the relations between the design parameters of the coarse model space and the fine model space. The above optimization result is equivalent to approaching the residual vector $f(x_f) = P(x_f) - x_c^*$ to 0.

2.3.2 *Application of Space Mapping in Insulation Prediction*

The basic idea of space mapping has some similarities with the research assumption of air insulation prediction:

1. The fine model and the coarse model involved in the space mapping method corresponds to two models of air gap discharge voltage prediction, one is the physical model based on gas discharge theories, the other is the insulation prediction model based on the grey correlation between the energy storage features and the discharge voltage.

The physical models simulate each discharge phase step-by-step. These models have rich physical connotations and they can reflect the development process of air discharge. However, the previous semi-empirical models usually based on many simplifications and assumptions, and the calculated results are not accurate enough. In addition, the direct numerical simulation based on micro models relies on numerical calculation ability and computer hardware performance, which is limited by the accuracy and efficiency of the numerical algorithms, therefore it is difficult to solve large-scale problems. Since the gas discharge theories are not yet perfect, the determination of some relevant parameters is based on some simplified assumptions, and the mathematical modeling of the dispersion and randomness of the discharge process is very difficult. Hence, it is still difficult to establish an accurate physical model at present. The physical significance of the insulation prediction model seems to be not enough, but the basis of this model is the correlation between the energy storage features and the discharge voltage, whose connotation coincides with the existing physical consensus. In addition, the computational efficiency of the insulation prediction model can be higher and it is easier to implement.

2. The design parameters of the two models in the space mapping technique respectively correspond to the key physical parameters of the physical model and the input energy storage features of the insulation prediction model.

The key physical parameters in the physical model can be concluded as follows:

- Streamer (corona) inception: electron collision ionization coefficient α, electron attachment coefficient η, electron diffusion coefficient D_e, electron drift velocity v_e, photoelectron emission coefficient γ_{ph}, photoelectron absorption coefficient μ, geometry factor g, streamer inception critical charge number N_{crit}, streamer inception critical electric field strength E_0, etc.
- Streamer propagation: space charge number in the streamer zone Q_C, electric field strength in the streamer channel E_s, streamer diameter D_s, streamer branch angle θ_s, streamer velocity v_s, etc.
- Leader inception: critical thermal ionization temperature, T_{crit}, etc.
- Leader propagation: charge density in the leader channel q_l, leader propagation velocity v_l, leader channel radius R_l, electric field strength in the leader channel E_l, etc.

The energy storage features of the insulation prediction model are as follows:

- Electric field features: electric field strength E, electric field energy W, electric field energy density W_d, electric field gradient E_g and the scale parameters characterizing the electric field inhomogeneity, etc.
- Voltage waveform features: impulse voltage amplitude U_m, voltage duration time (front time T_f, time to half-value T_2), voltage rise rate du/dt, voltage integral S, etc.
- Features of air dielectric properties: air pressure P, temperature T, and humidity H, etc.

It should be pointed out that some of the key physical parameters in physical models have not yet been unified, and the transformation mechanism of each discharge phase has not yet been completely clarified, hence it is difficult to form a fine physical model at present. However, the energy storage features of the insulation prediction model can be quantitatively characterized, and all the features are measurable and controllable.

3. The response of the two models in the space mapping technique corresponds to the output of the physical model and the insulation prediction model, namely, the discharge voltage.

In addition, there are also some differences between the basic ideas of the space mapping and air insulation prediction, mainly include the following two points:

1. The goal of space mapping method is to obtain the optimal parameter setting with the highest performance for a physical object, that is, the goal is to obtain the input parameters of the fine model. However, the goal of air insulation prediction is to acquire the discharge voltage under the known gap structure, applied voltage and atmospheric environment, that is to say, the goal is to obtain the output of the prediction model.
2. The key point of space mapping method is parameter extraction technology, that is, to establish the space mapping relation of the input parameters between the fine

model and the coarse model. However, the key point of air insulation prediction is to establish the mapping relation between the input parameters (energy storage features) and the output parameter (discharge voltage).

Based on the above similarities and differences, taking the idea of space mapping for reference, the air insulation prediction method can be constructed as follows:

1. As shown in Fig. 2.7, under the known gap structure, applied voltage and atmospheric environment, there are three methods to obtain the discharge voltage of an air gap. The first one is the discharge voltage measurement by experiments, the second one is the discharge voltage calculation by the fine physical model, with the known physical mechanisms of each discharge phase and the known physical parameters, the third one is the discharge voltage prediction by air insulation prediction model, established by mathematical algorithms with the extracted energy storage features. Among these 3 methods, the fine physical model is currently difficult to establish due to the imperfect gas discharge theories, the experiments are the main method to obtain the air gap discharge voltage, while the air insulation prediction model proposed in this book tries to provide an alternative of the fine physical model for air gap discharge voltage prediction.
2. The mapping relation between the test model and the insulation prediction model is established. Under the given applied voltage and atmospheric environment, the input parameters of the test model is the three-dimensional air gap structure, while the input parameters of the insulation prediction model are the electric field features, the structure space can be mapped to the electric field space, and their transformation relation can be expressed by the following mathematical formula:

$$A_{\mathrm{t}} \xrightarrow{P} A_{\mathrm{p}}$$
$$f_1(A_{\mathrm{t}}) = f_2(A_{\mathrm{p}}) \tag{2.35}$$

where A_{t} and A_{p} respectively represent the form of air gap in the structure space and the electric field space, f_i is the description function for air discharge problem in corresponding space, and P is the mapping function from the structure space to the electric field space. It should be noted that A_{t} can be characterized by geometric parameters for typical air gaps, such as the electrode radius and gap length, but for complex engineering gap arrangements, A_{t} used to accurately describe the three-dimensional structure is difficult to obtain. The electric field space can be used to describe the structures of any air gaps, but the electric field is of infinite dimensional distribution, and therefore A_{p} cannot fully represent the electric field distribution of an air gap. It is necessary to establish a feature set that can reflect the key information of the electric field distribution as much as possible. The mapping relation P from the structure space to the electric field space cannot be described by a simple mathematical function, but the electric field distribution of an air gap can be calculated by FEM, thus mapping A_{t} to A_{p}.

3. To solve the air discharge problem in the electric field space, the key is to establish the many-to-one mapping relation between the energy storage features and the

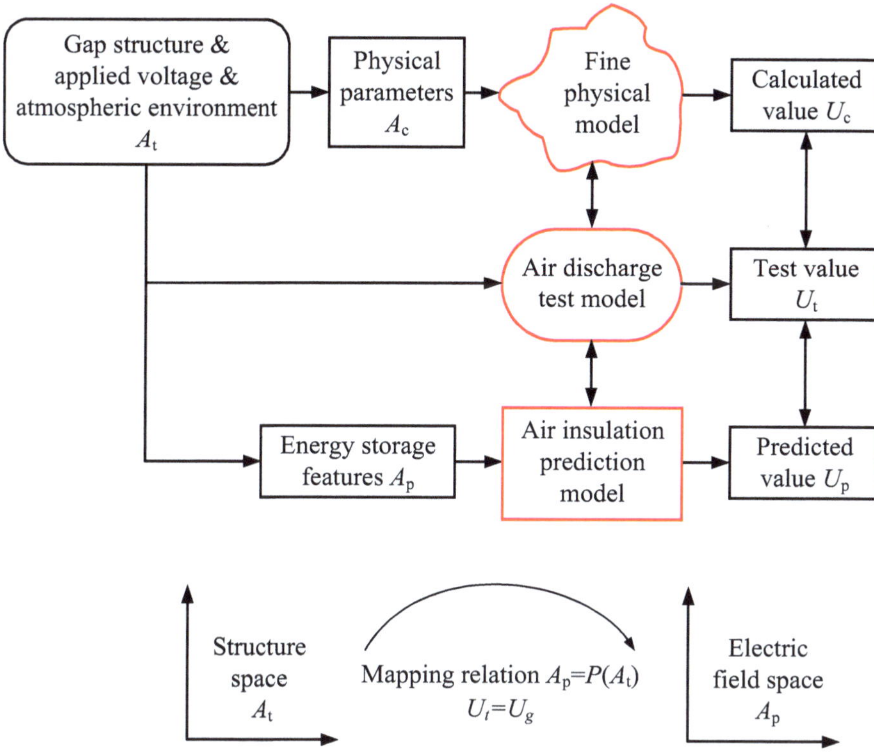

Fig. 2.7 Spatial mapping relation of air insulation prediction

discharge voltage, namely $U_p = f_2(A_p)$. However, the function relationship f_2 also cannot be described by simple mathematical functions, thus need to find other alternatives. Taking the space mapping technique for reference, whether the parameter extraction method meets the design requirements or not is verified by the fine model simulation, therefore some experimental results can also be introduced to the air insulation prediction model appropriately. That is to say, for a known gap structure, the structure space can be mapped to the electric field space to get the electric field features A_p, and the discharge voltage U_p can be obtained by experiments. If the energy storage features and discharge voltages of several air gaps are known, then the functional relationship between U_p and A_p can be fitted theoretically. However, A_p contains dozens of features, the relationships between these features and the discharge voltage U_p are not independent deterministic functions, but multi-dimensional nonlinear relationships. It is difficult to deduce widely applicable formulas, therefore the traditional regression analysis method is not suitable for accurate description of this problem. It is necessary to introduce some new mathematical methods.

4. The artificial intelligence or machine learning algorithms, like the artificial neural networks (ANN), support vector machine (SVM), have unique advantages

in dealing with complex multi-dimensional nonlinear problems. The key points are learning and generalization. The relationships and laws between the input parameters and the output parameters can be learned from the training samples, with the known input parameters of the test samples, the learned laws can be generalized to the test samples and therefore the output parameters can be predicted. By the artificial intelligence or machine learning algorithms, the correlations between the energy storage features and the discharge voltage can be established according to some known test data. Since the correlation cannot be described by deterministic functions, it can be viewed as a grey relation. Actually, this relation characterizes the intrinsic properties of the air discharge. Trained by known experimental data (training samples), the mathematical model with generalization ability can be used to predict the discharge voltages of other test samples.

2.4 Brief Summary

This chapter introduces the theoretical foundation of air insulation prediction. By analyzing the influence factors of air discharge, the feature set used to characterize the energy storage status of an air gap proposed from two aspects, including the electric field distribution and the impulse voltage waveform. The basic idea of space mapping and its application in air insulation prediction are illustrated. The relevant contents can be summarized as follows:

1. The electric field features are defined in three spatial positions, including the discharge channel, the shortest discharge path and the electrode surface, involving three spatial dimensions, namely, line, area and volume. The electric field features include the electric field strength, field energy, energy density, field gradient and some scale parameters characterizing the electric field inhomogeneity.
2. The voltage waveform features describe the switching and lightning impulse waveshapes from two aspects, including the global and the local characteristics, thus the features are divided into the basic features and the additional features. The impulse voltage waveform features include the impulse voltage amplitude, front time, time to half-value, voltage rise rate, voltage integral and other features characterizing the steepness, time and waveform integral features during the wave front.
3. For an air gap, its structural space can be mapped to the electric field space, in which the energy storage feature set can be constructed. The complex discharge process can be viewed as a grey box, and the many-to-one association mapping model between the energy storage features and the discharge voltage can be established directly by artificial intelligence or machine learning algorithms. Based on training and learning of the known data, it is possible to solve the air discharge problem.

References

1. Boggs S (2004) Analytical approach to breakdown under impulse conditions. IEEE Trans Dielectr Electr Insul 11(1):90–97
2. Xu XJ, Zhu DC (1996) Gas discharge physics. Fudan University Press, Shanghai
3. Ma NX (1998) Long gap discharge. China Electric Power Press, Beijing
4. Thione L, Pigini A, Allen NL et al (1992) Guidelines for the evaluation of the dielectric strength of external insulation. CIGRE Brochure, Paris, France
5. Shu SW, Ruan JJ, Huang DC et al (2015) Prediction for breakdown voltage of air gap based on electric field features and SVM. Proc CSEE 35(3):742–750
6. Gallimberti I (2015) A computer model for streamer propagation. J Phys D Appl Phys 5(12):2179–2189
7. Sarma MP, Janischewskyj W (1969) DC corona on smooth conductor in air: steady-state analysis of the ionisation layer. Proc IEE 116(1):161–166
8. Liao RJ, Wu FF, Liu KL et al (2015) Simulation of characteristics of electrons during a pulse cycle in bar-plate DC negative corona discharge. Trans China Electrotech Soc 30(10):319–329
9. IEC 60060-1 (2010) High-voltage test techniques—part 1: general definitions and test requirements
10. Harada T, Aihara Y, Aoshima Y (1973) Influence of switching impulse wave shape on flashover voltages of air gaps. IEEE Trans Power Appar Syst 92(3):1085–1093
11. Menemenlis C, Isaksson K (1974) The front shape of switching impulses and its effect on breakdown parameters. IEEE Trans Power Appar Syst 1974, 93(5):1380–1389
12. Bandler JW, Biernacki RM, Chen SH et al (1994) Space mapping technique for electromagnetic optimization. IEEE Trans Microw Theory Tech 42(12):2536–2544
13. Bakr MH, Bandler JW (2001) An introduction to the space mapping technique. Optim Eng 2(4):369–384
14. Deng JH (2007) Research of space mapping technology and its applications in LTCC circuit designs. Dissertation, University of Electronic Science and Technology

Chapter 3
Air Gap Discharge Voltage Prediction Model

3.1 Algorithm Selection of Prediction Model

Artificial intelligence or machine learning algorithms, such as ANN and SVM, have good performance in dealing with multi-dimensional nonlinear problems. They have been widely applied in many areas of power system research. However, these algorithms are seldom applied in discharge voltage prediction and external insulation design of transmission projects. Only a few scholars carried out investigations on the ANN and SVM regression models about the influences of atmospheric conditions on air gap breakdown voltage. The selection of appropriate intelligent algorithms is of great importance to ensure the effectiveness and accuracy of air gap insulation prediction model.

3.1.1 Applications of Artificial Intelligence Algorithms

In recent twenty years, the idea of predicting air gap breakdown voltages by some artificial intelligence algorithms has always attracted interests of researchers. Various artificial neural networks, fuzzy logic, and SVM have already been applied in correlation factor analysis of air discharge and breakdown voltage prediction.

A few researchers use artificial intelligence algorithms to analyze the physical process and physical quantities of air discharge. For example, using streamer inception criterion to establish inference data base, the fuzzy logic technique was introduced in [1, 2] to predict the positive DC breakdown voltages of point-plane air gaps. The input parameters are the electrode radius, gap distance and the applied voltage, while the outputs are 4 discharge steps including the Townsend avalanche growth, corona, streamer and breakdown. The neural network was used in [3] to analyze the breakdown process for plane-rod gap discharge, and the results indicate that the physical quantities measured at early stages of the negative switching impulse discharge test have some relevance with the discharge results, namely, breakdown or withstanding.

Z. Qiu et al., *Air Insulation Prediction Theory and Applications*, Power Systems,
https://doi.org/10.1007/978-981-10-5163-0_3

The radial basis function Gaussian network (RBFG) and self-organization neural network were applied in [4, 5] to predict the breakdown voltages of point-barrier-plane air gaps. The neural network was trained by a set of samples with the form (y_i, y_{i+1}), $i = 1, 2, \ldots, n-1$, and the weights of the network were updated after this training so as to predict a future value y_{n+1}. A prediction model based on support vector regression (SVR) was proposed in [6] to predict the corona onset voltage of AC conductor. The input parameters were relative air density, humidity, and roughness coefficient, while the output was the corona onset voltage. The predicted results consist with the test results. The BP neural network was used in [7] to predict the AC breakdown voltages of short rod-plane air gaps under rain conditions. The gap length, rainfall intensity, rain conductivity, temperature and absolute humidity were taken as the input parameters and the breakdown voltage was the output. The results showed that training sample data play an essential role in the prediction ability of ANN and the predicted results for the sample out of the training range are of unacceptable error. The BP neural network, RBF neural network and Chebyshev neural network were applied in [8–10] for breakdown voltage prediction under different atmospheric conditions. The grey correlation analysis method was used to study the impact of atmospheric parameters on air gap discharge voltage. Taking the atmospheric parameters like air pressure, temperature, humidity and wind speed as inputs, the sphere gap breakdown voltages under different atmospheric conditions were predicted accurately, both in artificial climate chamber and natural environment. The breakdown voltages of imitated air gaps were predicted by BP neural network in [11]. Trained by the experimental data of air gaps with several gap lengths, it can predict air gap breakdown voltages with other gap lengths, under the same air pressure. Also, according to the experimental data under different air pressures, the network can predict the breakdown voltages under other air pressure.

The above research results preliminary verified the feasibility of artificial intelligence and machine learning algorithms in discharge voltage prediction of air gaps. However, these researches are still not deep enough. There are two main problems. Firstly, most of the existing researches predict the breakdown voltages of the same air gap under different atmospheric parameters. These prediction models do not consider the gap structure. How to realize discharge voltage prediction of different air gap configurations still needs further studies. Secondly, most of the reported ANN and SVM models need to be trained by many test data, which is against to reduce experimental work. For long air gaps and complex engineering gap configurations, it is extremely expensive to obtain their experimental discharge voltages. How to achieve air insulation prediction in small sample conditions also needs more studies.

3.1.2 Basis for Algorithm Selection

Artificial intelligence algorithms cannot be optionally applied. In order to mine effective information from the known data and realize accurate prediction of the unknown data, various factors should be analyzed systematically, including the nature of the problem, the sample size, the feature data, the algorithm design and the parameter selection, etc. Aiming at air insulation prediction, it should be clarified that the algorithms like ANN or SVM are still in essence mathematical models. The mapping relations between the input features and the output quantities can be obtained by data training, but they cannot be expressed by explicit mathematical functions. SVM is based on statistical learning theory, and it has many unique advantages in dealing with small sample and multi-dimensional nonlinear problems. The existing studies also demonstrate that its prediction performance is better than traditional ANN models. The objective of air insulation prediction is to obtain air gap discharge voltages under actual spatial electric field distribution, applied voltage and atmospheric conditions, by the method of small-sample machine learning. In other words, it is to establish the mapping relations between the above influencing factors and the discharge voltage.

SVM is a new machine learning algorithm developed on the basis of statistical learning theory. Compared with other traditional artificial intelligence methods, SVM has the following advantages [12].

1. SVM is based on strict statistical learning theory and structural risk minimization (SRM) principle, instead of empirical risk minimization (ERM) principle, which simultaneously minimizes the empirical risk and confidence interval of training samples so as to ensure the generalization performance of learning. SVM overcomes the over fitting problem in traditional neural networks resulted from function estimation only depending on empirical risk minimization. Therefore, SVM has unique advantages in solving small sample learning problems, without using the asymptotic conditions when the sample tends to infinity.
2. SVM transforms the learning problem into a convex quadratic programming optimization problem and it can theoretically obtain the global optimal solution. Hence, SVM solves the local extremum problem which is unavoidable for traditional neural network method.
3. By introducing kernel functions, SVM can transform the problem from a nonlinear space into a high-dimensional linear space. This makes the complexity of the algorithm independent of the sample dimension, and skillfully overcomes the "curse of dimensionality" caused by rapid increase of data size when mapping from low-dimensional nonlinear space to high-dimensional linear space.

3.2 Fundamental Theory of SVM

Currently, air gap discharge voltages are mainly measured by experiments. For large scale gaps with complex structures, it is usually difficult to obtain the test data. Hence, the air gap discharge voltages belong to small sample data. SVM is a machine learning algorithm put forward by V. N. Vapnik based on VC dimension in statistical learning theory [13] and structural risk minimization principle. It has good generalization ability for pattern recognition of small sample data.

3.2.1 Statistical Learning Theory

The VC dimension in statistical learning theory and the SRM principle are the theoretical foundations of SVM. In order to derive the VC dimension theory and the SRM principle, the concepts of generalization ability, loss function, expected risk and empirical risk [14, 15] are firstly introduced briefly.

1. Generalization ability

The solution of the classification problem is to find an optimal decision function $f(x)$ according to the training sample set.

$$\begin{aligned} & T = \{(x_1, y_1), \ldots, (x_l, y_l)\} \in \{X \times Y\}^l \\ & x_i \in X \subset \mathbf{R}^n, y_i \in Y = \{-1, 1\}, \quad i = 1, 2, \ldots, l \end{aligned} \tag{3.1}$$

$f(x)$ is a function defined on X and its value is -1 or 1.

$$f : X(X \subset \mathbf{R}^n) \rightarrow Y = \{-1, 1\} \tag{3.2}$$

The so-called "generalization" is to infer the corresponding output y for a new input x according to $y = f(x)$, on condition that the decision function $f(x)$ is determined. The generalization ability is used to evaluate the merits of the "generalization". If an algorithm is of strong generalization ability, it means that the classifier has high classification accuracy for test samples when it is trained by limited training samples [16].

2. Expected risk and loss function

The expected risk involves probability distribution and loss function. They are introduced as follows.

- Probability distribution. Set $X \subset \mathbf{R}^n$, $Y = \{-1, 1\}$, and (x, y) is the random vector, where $x = ([x]_1, \ldots, [x]_n)^{\mathrm{T}} \in X$, $y \in Y$, the function $P(\bar{x}, \bar{y}) = P(x \leq \bar{x}, y \leq \bar{y})$ is the probability distribution on $X \times Y$. Here, $\bar{x} = ([\bar{x}]_1, \ldots, [\bar{x}]_n)^{\mathrm{T}} \in X$, $\bar{y} \in Y$. $P(x \leq \bar{x}, y \leq \bar{y})$ is the probability of the simultaneous occurrence of event "$x \leq \bar{x}$" and "$y \leq \bar{y}$". "$x \leq \bar{x}$" means "$[x]_1 \leq [\bar{x}]_1, \ldots, [x]_n \leq [\bar{x}]_n$". The function $P(\bar{x}, \bar{y})$ is also usually denoted as $P(x, y)$.
- Loss function. Set $X \subset \mathbf{R}^n$, $Y = \{-1, 1\}$, a triple $(x, y, f(x)) \in X \times Y \times Y$ is defined, where x is a pattern, y is a observed value, and $f(x)$ is a predicted value. If the mapping c: $X \times Y \times Y \rightarrow [0, \infty)$ bring $c(x, y, y) = 0$ for any $x \in X$, $y \in Y$, then c is called a loss function.

In short, the loss function $c(x, y, y)$ is a function of $c(x, y, f(x)) = 0$ when $f(x) = y$. It means that the loss value is zero when the prediction is accurate and the loss value is not zero when the prediction error reaches a certain degree.

- Expected risk. Set $P(x, y)$ as the probability distribution on $X \times Y$, c is the given loss function and $f(x)$ is the decision function defined by Eq. 3.2, then the expected risk of the decision function $f(x)$ is the Riemann-Stieltjes integral of the loss function with respect to the probability distribution $P(x, y)$.

$$R[f] \triangleq \mathbf{E}[c(x, y, f(x))] = \int_{X \times Y} c(x, y, f(x)) \mathrm{d}P(x, y)$$
$$= \int_X c(x, -1, f(x)) \mathrm{d}P(x, -1) + \int_X c(x, -1, f(x)) \mathrm{d}(P(x, 1) - P(x, -1)) \tag{3.3}$$

The value $c(x, y, f(x))$ of the loss function is the quantitative index to evaluate the performance of the decision function $f(x)$ at the input x, while the expected risk is the quantitative index to evaluate the overall effect of the decision function $f(x)$.

3. ERM principle

When the training sample set defined by Eq. 3.1 is known, set these sample points are independent and identically distributed according to the probability distribution $P(x, y)$ on a certain unknown $X \times Y$, and set the loss function $c(x, y, f(x))$ is given, then the classification problem is to find a decision function $f(x)$ that minimizes the expected risk $R[f]$. However, the known condition is only the training sample set T composed of finite points generated by a certain probability distribution $P(x, y)$ and the concrete form of $P(x, y)$ is unknown. Hence, the expected risk of a decision function $f(x)$ is incalculable. It is necessary to find a quantitative index to replace the expected risk, which can be calculated and can also reflect the quality of a decision function to a certain extent. So, the following quantitative index is introduced as "empirical risk".

Given the training sample set defined by Eq. 3.1 and the loss function $c(x, y, f(x))$, the empirical risk of the decision function $f(x)$ is

$$R_{\mathrm{emp}}[f] = \frac{1}{l} \sum_{i=1}^{l} c(x_i, y_i, f(x_i)) \tag{3.4}$$

The ERM principle to solve classification problems is to select a decision function candidate set F, consisting of several functions defined on $\mathbf{R}^n$ and their values belong to $Y = \{-1, 1\}$, and select the function f minimizing the empirical risk in the set F as the decision function. When the number of training samples l is enough, the empirical risk can represent the expected risk in general. However, when the number of training samples l is not enough, there is a large difference between the empirical risk and the expected risk. This is the limitation of the ERM principle.

4. VC dimension

The ERM principle belongs to the category of traditional statistical theory, and the decision function candidate set F is determined in advance, while the statistical learning theory takes into account the effect of set F on the expected risk. The VC dimension proposed by V. N. Vapnik and A. Chervonenkis is a quantitative index to describe the size of the set F. The concept of VC dimension is based on "scatter" of the point set by F.

Set F as a decision function candidate set consisting of several functions defined on $X \subset \mathbf{R}^n$ and their values are -1 or 1, and $Z_m = \{x_1, \ldots, x_m\}$ is a set consisting of m points in X. When f takes all the possible decision functions in F, the resulted m-dimensional vector is $(f(x_1), \ldots, f(x_m))$. Defining $N(F, Z_m)$ as the number of different vectors in the above m-dimensional vector, if $N(F, Z_m) = 2^{m}$, then it is called that Z_m is scattered by F.

If there is a set Z_m consisting of m points that can be scattered by F, and any set Z_{m+1} consisting of $m + 1$ points cannot be scattered by F, then the VC dimension of F is m. For any positive integer m, if there is a set Z_m consisting of m points that can be scattered by F, then the VC dimension of F is infinite. Thus, the VC dimension of F is the maximum number of the points in X that can be scattered by F. The VC dimension reflects the learning ability of the function set. The larger the VC dimension, the more complex the learning machine. Hence, the VC dimension is an evaluation of the complexity of the learning machine [16].

5. SRM principle

When the number of training samples is limited, it is impossible to use empirical risk to approximate the expected risk. It needs to introduce the concept of structural risk. Firstly, the following theorems are introduced.

Set h as the VC dimension of F, if

$$l > h, \quad h\left(\ln \frac{2l}{h} + 1\right) + \ln \frac{4}{\delta} \geq \frac{1}{4} \tag{3.5}$$

For arbitrary probability distribution $P(x, y)$ and $\delta \in (0, 1]$, the arbitrary decision function f in F can make the following inequation hold with the probability $1 - \delta$ at least.

$$R[f] \leq R_{\mathrm{emp}}[f] + \sqrt{\frac{8}{l}\left[h\left(\ln \frac{2l}{h} + 1\right) + \ln \frac{4}{\delta}\right]} \tag{3.6}$$

The first item $R_{\mathrm{emp}}[f]$ at the right of the Eq. 3.6 is the empirical risk, and the second item

$$\varphi(h, l, \delta) = \sqrt{\frac{8}{l}\left[h\left(\ln \frac{2l}{h} + 1\right) + \ln \frac{4}{\delta}\right]} \tag{3.7}$$

is called the confidence interval. The sum of these two items is called structural risk.

The above theorem shows that the structural risk is an upper bound of the expected risk of $R[f]$ and the confidence interval is an estimate of the difference between the expected risk and the empirical risk, which is a decreasing function of the training

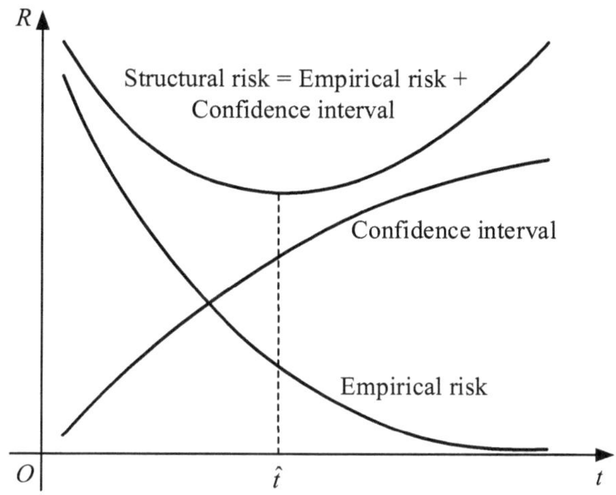

Fig. 3.1 Structural risk, empirical risk and confidence interval

sample number l and tends to zero when $l \to \infty$. When the number of training samples is large, the confidence interval is small, and the empirical risk can replace the expected risk. When the number of training samples is small, the confidence interval should be taken into account. In addition, the confidence interval is related to the VC dimension h that describes the size of the decision function candidate set F, and the effect of the size of F on the expected risk should be considered.

As shown in Fig. 3.1, the horizontal axis describes the size of F, and the vertical axis describes the value corresponding to the right of the Eq. 3.6. On one hand, it can be seen that the candidate function increases and the empirical risk decreases with an increased size of the set F. On the other hand, when the size of F increases, its VC dimension h and the confidence interval will increase. Hence, in order to minimize the structural risk, the influences of F on the empirical risk and the confidence interval should be together taken into account and the set F with an appropriate size should be selected.

The SRM principle to solve classification problems is to select a decision function set $F(t)$ depended on the parameter t, namely, $F(t_1) \subset F(t_2)$, $\forall\, t_1 < t_2$, a function f^t that minimizes the empirical risk can be found in the set $F(t)$ for every t, and there is a corresponding value of the structural risk. The minimum $t = \hat{t}$ that minimizes the structural risk is selected, and the corresponding function $f^{\hat{t}}$ is taken as the decision function.

3.2.2 *Support Vector Classifier*

SVM was primarily applied to solve the binary classification problem. The basic theory and the implementation process of the support vector classifier (SVC) are briefly introduced in this section.

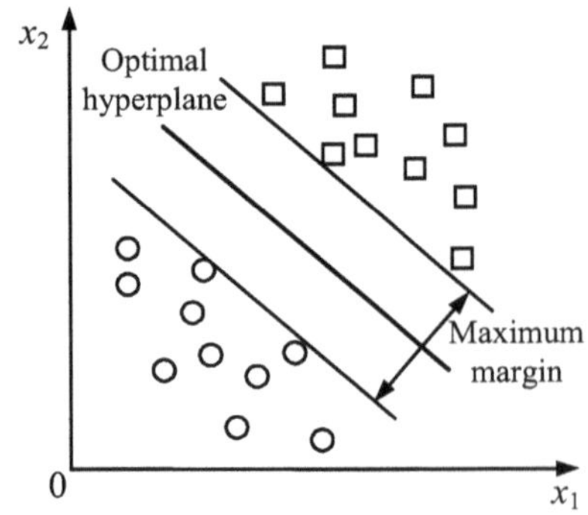

Fig. 3.2 Schematic diagram of the maximum margin (linearly separable)

Given a training sample set shown in Eq. 3.1, if there is an optimal separating hyperplane $w \cdot x + b = 0$ that can divide the training sample set into two groups without error, this is called linearly separable condition. As shown in Fig. 3.2, based on the maximum margin principle, the above linearly separable problem should maximize the margin between two support hyperplanes and the optimization problem is the convex quadratic programming problem for variables w and b.

$$\begin{cases} \min\limits_{w,b} & \dfrac{1}{2}\|w\|^2 \\ \text{s.t.} & y_i(w \cdot x_i + b) \geq 1, \quad i = 1, 2, \ldots, l \end{cases} \tag{3.8}$$

In order to make the solution of the primal optimization problem (3.8) more simple and practicable, the Lagrange function is introduced.

$$L(w, b, \alpha) = \frac{1}{2}\|w\|^2 - \sum_{i=1}^{n} \alpha_i \{y_i[(w \cdot x_i) + b] - 1\} \tag{3.9}$$

where $\alpha = (\alpha_1, \ldots, \alpha_n)^{\mathrm{T}}$ is the Lagrange multiplier vector. By calculating the partial derivatives for w and b, and making them equal to 0 according to the extremum condition, the primal problem (3.8) can be transformed to the following dual problem.

$$\begin{cases} \max\limits_{\alpha} & -\dfrac{1}{2}\sum\limits_{i=1}^{n}\sum\limits_{j=1}^{n} \alpha_i \alpha_j y_i y_j (x_i \cdot x_j) + \sum\limits_{j=1}^{n} \alpha_j \\ \text{s.t.} & \sum\limits_{i=1}^{n} \alpha_i y_i = 0, \quad \alpha_i \geq 0, \ i = 1, 2, \ldots, l \end{cases} \tag{3.10}$$

The formulated support vector machine for linearly separable condition is called the hard-margin support vector machine. The primal problem (3.8) is a convex quadratic programming problem with linear inequality constraints, so its dual problem must be solvable. Set $\alpha^* = (\alpha_1^*, \ldots, \alpha_n^*)^{\mathrm{T}}$ as the arbitrary solution of Eq. (3.10), then the solution of the primal problem is

$$w^* = \sum_{i=1}^{n} \alpha_i^* y_i x_i \tag{3.11}$$

Selecting a positive component α_j^* of α^*, b^* can be calculated by

$$b^* = y_j - \sum_{i=1}^{n} \alpha_i^* y_i (x_i \cdot x_j) \tag{3.12}$$

Substituting Eq. 3.11 into the hyperplane equation $w^* \cdot x + b^* = 0$, the decision function can be obtained.

$$f(x) = \mathrm{sgn}\left(\sum_{i=1}^{n} \alpha_i^* y_i (x_i \cdot x) + b^*\right) \tag{3.13}$$

For general classification problems, there may be linearly inseparable conditions. Therefore the above separating hyperplane does not exist, and the hard-margin support vector machine is unsolvable. At this moment, the training samples which do not satisfy the constraint $y_i(w \cdot x_i + b) \geq 1$ must be allowable, which means to "soften" the requirement of the separating hyperplane and allow the existence of wrongly classified sample. This is called approximately linearly separable condition. To allow inseparability, the slack variable ξ_i (≥ 0) and the penalty factor C are introduced. As shown in Fig. 3.3, if $0 < \xi_i < 1$, the data do not have the maximum margin but are still correctly classified. If $\xi_i \geq 1$, like ξ_j in Fig. 3.3, the data are misclassified by the optimal separating hyperplane. For approximately linearly separable conditions, the primal problem (3.8) can be changed into the following primal optimization problem.

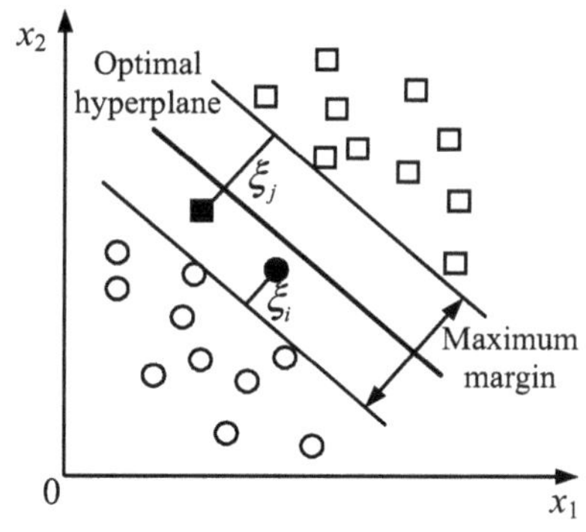

Fig. 3.3 Approximately linearly separable condition

$$\begin{cases} \min\limits_{w,b,\xi} & \dfrac{1}{2}\|w\|^2 + C\sum\limits_{i=1}^{n}\xi_i \\ \text{s.t.} & y_i(w \cdot x_i + b) \geq 1 - \xi_i, \quad \xi_i \geq 0, \ i = 1, 2, \ldots, l \end{cases} \tag{3.14}$$

where $\xi = (\xi_1, \ldots, \xi_n)^{\mathrm{T}}$, $C > 0$.

Similarly, the following Lagrange function is introduced.

$$L(w, b, \xi, \alpha, \beta) = \frac{1}{2}\|w\|^2 + C\sum_{i=1}^{n}\xi_i - \sum_{i=1}^{n}\alpha_i\{y_i[(w \cdot x_i) + b] - 1 + \xi_i\} - \sum_{i=1}^{n}\beta_i\xi_i \tag{3.15}$$

where $\alpha = (\alpha_1, \ldots, \alpha_n)^{\mathrm{T}}$ and $\beta = (\beta_1, \ldots, \beta_n)^{\mathrm{T}}$ are the Lagrange multiplier vectors. By solving the partial derivatives of Eq. 3.15 for w, b and ξ_i, according to the extremum conditions, the primal optimization problem (3.14) can be transformed to the following dual problems:

$$\begin{cases} \max\limits_{\alpha,\beta} & -\dfrac{1}{2}\sum\limits_{i=1}^{n}\sum\limits_{j=1}^{n}\alpha_i\alpha_j y_i y_j (x_i \cdot x_j) + \sum\limits_{j=1}^{n}\alpha_j \\ \text{s.t.} & \sum\limits_{i=1}^{n}\alpha_i y_i = 0 \\ & C - \alpha_i - \beta_i = 0, \quad i = 1, 2, \ldots, l \\ & \alpha_i \geq 0, \beta_i \geq 0, \quad i = 1, 2, \ldots, l \end{cases} \tag{3.16}$$

This support vector machine for approximately linearly separable condition is called the soft-margin support vector machine. In order to simplify the solution of the dual problem, the variable β can be eliminated by equality constraint $C - \alpha_i - \beta_i = 0$, $i = 1, 2, \ldots, l$, and therefore it can be transformed to the problem only containing the variable α. Similarly, by solving the dual problem, the solution of the primal optimization problem can be obtained and thus obtain the decision function (3.13).

In actual situations, the training samples are usually linearly inseparable. Therefore, the original input data should be mapped into a high dimensional linear space by a kernel function. Then in this high dimensional space, the sample data can be classified by the optimal separating hyperplane. This high dimensional space is called the feature space, as shown in Fig. 3.4. The form of the kernel function is

$$K(x_i, x_j) = (\phi(x_i) \cdot \phi(x_j)) \tag{3.17}$$

where $(\cdot)$ represents the inner product.

After nonlinear transformation by kernel function, the Eq. 3.10 can be transformed to

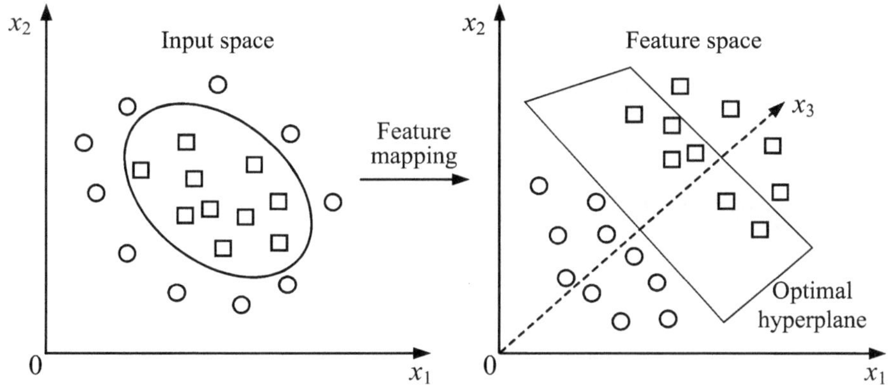

Fig. 3.4 Feature mapping

$$\begin{cases} \max\limits_{\alpha} & -\frac{1}{2}\sum_{i=1}^{n}\sum_{j=1}^{n}\alpha_i\alpha_j y_i y_j K(x_i, x_j) + \sum_{j=1}^{n}\alpha_j \\ \text{s.t.} & \sum_{i=1}^{n}\alpha_i y_i = 0, 0 \le \alpha_i \le C, \quad i = 1, 2, \ldots, n \end{cases} \tag{3.18}$$

By solving the above optimization problem, the decision function can be obtained.

Kernel technique is one of the most important problems in SVM, and the common kernel functions are as follows.

1. Linear kernel function

$$K(x_i, x_j) = x_i^{\mathrm{T}} x_j \tag{3.19}$$

2. Polynomial kernel function

$$K(x_i, x_j) = (\gamma x_i^{\mathrm{T}} x_j + r)^p, \gamma > 0 \tag{3.20}$$

3. Radial basis function (RBF) kernel function

$$K(x_i, x_j) = \exp(-\gamma \left\| x_i - x_j \right\|^2), \gamma > 0 \tag{3.21}$$

4. Sigmoid kernel function

$$K(x_i, x_j) = \tanh(v x_i^{\mathrm{T}} x_j + c) \tag{3.22}$$

where v is the first order constant, and c is the bias term.

RBF kernel function has better generalization ability, and it is the most widely used kernel function at present. In Eq. 3.21, γ is the kernel parameter. In this book, RBF kernel function is selected to be used in air gap discharge voltage prediction.

3.3 Parameter Optimization Methods

From the above analysis, it can be seen that the penalty coefficient C and the kernel parameter γ determine the classification and recognition performance of the SVM classifier. In order to obtain the optimal parameters, the cross validation (CV) is introduced, while C and γ are taken as the optimization variables. The improved grid search (GS) algorithm, genetic algorithm (GA) and particle swarm optimization (PSO) algorithm are respectively applied to optimize the SVM parameters.

3.3.1 Cross Validation

Cross validation was first proposed by Stone [17], which is a statistical analysis method applied to verify the performance of the classifier and used to analyze its generalization ability. The basic idea is as follows. The original sample data are divided into a training sample set and a verification sample set. Firstly, the training sample set is used to train the classifier. Then the trained prediction model is tested by the verification sample set, and the classification accuracy of the test is taken as an index to evaluate the performance of the classifier. The common CV methods include K-fold cross validation (K-CV) and leave-one-out cross validation (LOO-CV).

K-CV is to divide the original sample data into K groups. Taking each subset data as the verification sample set for one time and the other $K - 1$ subsets as the training sample set, K times of the test is required. The average value of the classification accuracy of the K tests is taken as the performance index of the SVM classifier. Generally speaking, $K \geq 2$, and the actual test process is usually started from $K =$ 3. K-CV can effectively avoid under fitting and over fitting, and the test results also have high persuasion.

Given an original sample data with N samples, LOO-CV takes each sample data separately as the verification sample and the other $N - 1$ sample data as the training sample set, thus obtain N models. The average value of the classification accuracy of the verification samples obtained by N tests is taken as the performance index of SVM classifier. LOO-CV is actually N-CV, but it uses almost all samples in the original sample data to train the model, and the evaluated results are more reliable. Meanwhile, the test process can be repeated because the sample data are not affected by random factors [18]. However, LOO-CV needs to establish N verification models, and the computational process is complex and time-consuming. Hence, K-CV is applied to optimize SVM parameters in this book.

3.3.2 Grid Search Algorithm

GS algorithm is the most basic parameter optimization algorithms. Its realization process is: (1) The search range and the step size of the penalty coefficient C and the kernel parameter γ are respectively set for grid meshing. (2) The values of C and γ in each grid interval are substituted into the SVM model one by one to verify its classification accuracy in the sense of K-CV. (3) The values of the SVM classification accuracy under each (C, γ) are compared with each other, and that with the highest classification accuracy is taken as the optimal parameters of the SVM model. Therefore, when the search range and the step size are set reasonably, the GS algorithm can theoretically find the optimal solution of the parameters. But it is often with large amounts of computation and time-consuming.

In order to give consideration to both the searching precision and efficiency, the coarse and fine grid search can be performed to find the optimal parameter group (C, γ), that is to make proper improvements of the GS algorithm [19]. Firstly, the large search range and step size are set for coarse grid search of the optimal parameters. Then according to the optimal parameters determined by coarse grid search, a smaller search range is defined nearby and the fine grid search is carried out with a smaller step size, thus to obtain the final optimal parameter group (C, γ).

The improved GS algorithm can greatly reduce the searching time of traditional GS algorithm with large search range and small step size, but it relies on experience to a certain extent in the search range set of the fine grid search, which brings the possibility of local optimization and reduces the classification accuracy of SVM to a certain degree.

3.3.3 Genetic Algorithm

GA is a global random search and optimization algorithm developed on the basis of evolution theory and genetics. The optimization problem is simulated as the survival of the fittest in a group. Starting from any one initial population, by the operation of random selection, crossover and mutation, the reproduction and evolution from generation to generation of the group will finally converge to an individual with the best fitness to environment [20]. The implementation process of the optimization of SVM parameters by GA is shown in Fig. 3.5, and the details are as follows:

1. Several groups of SVM parameters are generated randomly in the given search range, and they are binary coded to generate the initial population.
2. The fitness function is determined by the classification accuracy of the training samples under K-CV. The fitness function values of the individuals belong to this population are calculated to evaluate their merits. The smaller the fitness, the larger the error.
3. The individuals with high fitness are selected and directly inherited to the next generation, so as to make them enter the next iterative process.

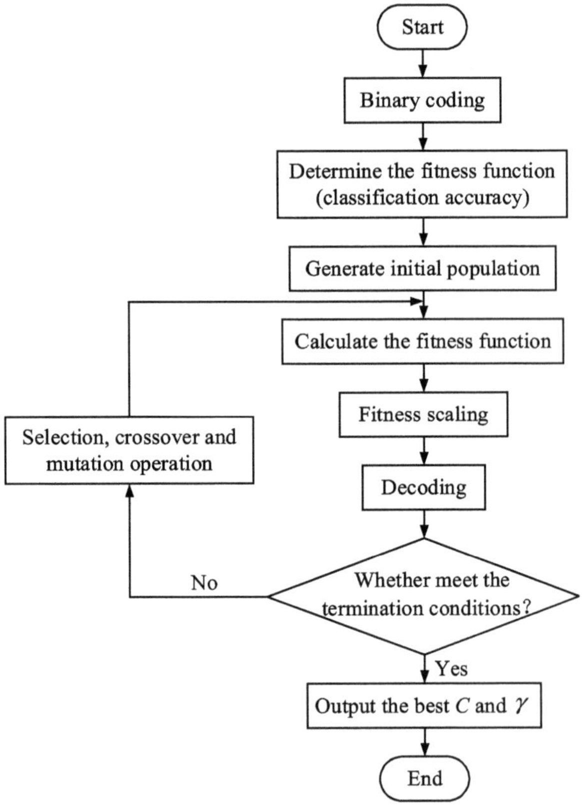

Fig. 3.5 Flow chart of parameter optimization by GA

4. The crossover and mutation operations are conducted for the current population to generate the next generation. By repeating the step 2, the SVM parameters generated by the initial population will evolve continuously until the terminal conditions of the training objective are satisfied. Then the optimal solution is obtained and the best C and γ will be output by decoding.

3.3.4 Particle Swarm Optimization Algorithm

PSO algorithm is a global optimization algorithm based on swarm intelligence theory. The swarm intelligence generated by cooperation and competition among particles in a population is used to guide the optimizing search. Compared with GA, PSO algorithm also has the characteristics of evolutionary computation and global searching strategy based on population. But its operation is relatively simple and it uses the velocity-displacement model to avoid the complex genetic manipulation in GA such as selection, crossover, mutation, and so on. The basic idea of PSO algorithm is introduced as follows [20]:

1. Taking the solution of the optimization problem as "particle" in the search space, firstly an initial population containing several particles is randomly initialized in the feasible solution space, and each particle in the population represents a feasible solution of the optimization problem.
2. A fitness function is constructed, and a fitness value is determined for each particle. The minimum value of the fitness function and the maximum number of population iterations are set as the convergence criteria.
3. Each particle will move with its own velocity and direction in the solution space, and the search usually follows the current optimal particle. In each iteration, the particle will adjust and update its flight trajectory according to the optimal solution found by itself (its own extremum) and that found by the whole population (global extremum), so as to search the optimal solution. The flight trajectory of the particle has memory characteristics and it can simultaneously memorize the location and velocity information. Hence, it can find the final optimal solution quickly.

In PSO algorithm, the fitness function is also the classification accuracy of training samples in K-CV sense. The implementation process of the optimization of SVM parameters by PSO algorithm is shown in Fig. 3.6.

In the above 3 parameter optimization algorithms, the improved GS algorithm is the simplest, and its optimization results are repeatable, but the results may be the local optimal solution. GA and PSO algorithm can find the global optimal solution in theory, but both of them are heuristic algorithms. The results of each optimization are different, and therefore the optimization is unstable. In general, these three optimization algorithms have their own advantages and disadvantages. In practical applications, parameter optimization algorithms should be selected reasonably according to the search efficiency and precision.

3.4 Feature Dimension Reduction Methods

The air gap energy storage features defined in this book are not entirely independent of each other. The correlation between features may lead to information redundancy, and it may cause insufficient information when the number of features is not enough. These two conditions can both affect the classification and recognition ability of the SVM classifier. Therefore, the correlation analysis method and the principal component analysis method are used for dimension reduction of the energy storage features, and the influences of different dimensions on the predicted results are analyzed.

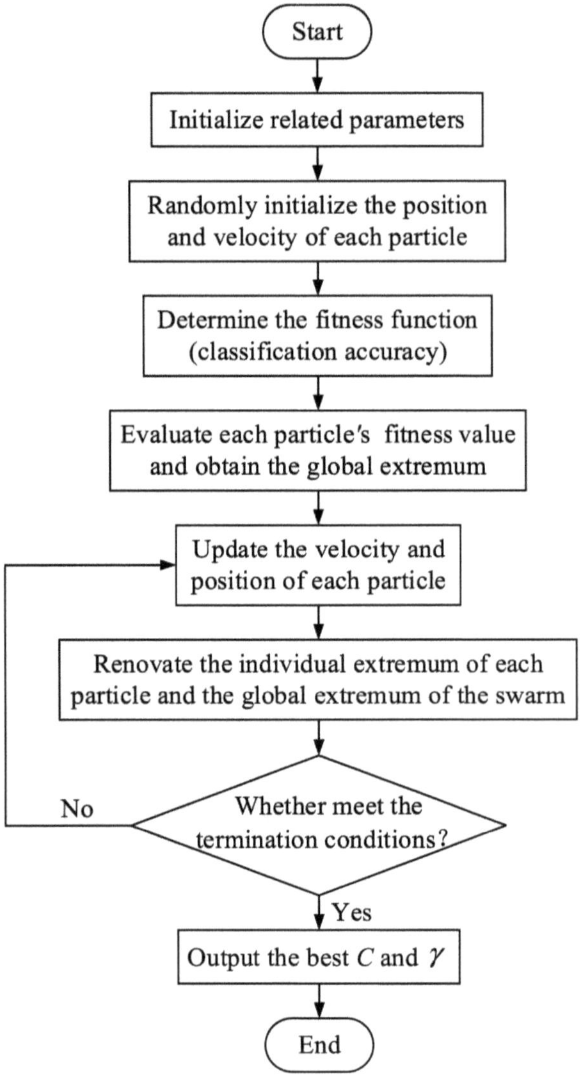

Fig. 3.6 Flow chart of parameter optimization by PSO algorithm

3.4.1 Normalization Processing

Since the physical meaning of each feature is different, in order to avoid the effect of different order of magnitudes and dimension on the predicted results, each feature should be normalized to the interval [0, 1] before feature dimension reduction and input to the SVM model. The normalization method is

$$x_i' = \frac{x_i - x_{i\min}}{x_{i\max} - x_{i\min}} \tag{3.23}$$

where x'_i is the normalized value of a certain feature x_i, $x_{i\max}$ and $x_{i\min}$ are the maximum and minimum values of x_i.

3.4.2 Correlation Analysis Method

Correlation analysis is a common feature dimension reduction method. By calculating the correlation coefficients between the features and the output target value and the cross correlation coefficients between different features, the features displayed a strong correlation with the target value are selected, while those displayed a weak correlation with the target value, or with a strong cross correlation are rejected. The correlation degree between two features is usually evaluated by Pearson correlation coefficient r. Its mathematical formula is

$$r = \frac{\sum (x_i - \bar{x})(y_i - \bar{y})}{\sqrt{\sum (x_i - \bar{x})^2 (y_i - \bar{y})^2}} = \frac{n\sum_{i=1}^{n} x_i y_i - \sum_{i=1}^{n} x_i \cdot \sum_{i=1}^{n} y_i}{\sqrt{n\sum_{i=1}^{n} x_i^2 - \left(\sum_{i=1}^{n} x_i\right)^2} \cdot \sqrt{n\sum_{i=1}^{n} y_i^2 - \left(\sum_{i=1}^{n} y_i\right)^2}} \tag{3.24}$$

where x_i and y_i are the ith values of any two features x and y, while $\bar{x}$ and $\bar{y}$ are their average values.

The correlation degree between two features is usually divided into 4 grades based on the absolute value of the Pearson correlation coefficient r. $0 < |r| \leq 0.3$ is basic uncorrelation, $0.3 < |r| \leq 0.5$ is low correlation, $0.5 < |r| \leq 0.8$ is significant correlation, and $0.8 < |r| \leq 1$ is high correlation.

3.4.3 Principal Component Analysis Method

Principal component analysis (PCA) is a multivariate statistical analysis method for feature extraction. The basic principle is to recombine the original multiple features with strong correlations and transform them to a few unrelated aggregate variables called principal component. These principal components are used to replace the original features as input parameters of the prediction model. The detailed procedures of PCA are as follows.

1. The original feature data are normalized by Eq. 3.23.
2. The correlation coefficient matrix $\boldsymbol{R} = (r_{ij})_{p\times p}$ is calculated. r_{ij} is the correlation coefficient of any two features x and y, which can be calculated by Eq. 3.24. p is the feature dimension.
3. The characteristic roots and eigenvectors of $\boldsymbol{R}$ are calculated. The characteristic root λ can be calculated by solving the characteristic equation $|\boldsymbol{R} - \lambda \mathbf{I}| = 0$, and they can be arranged in order of size $\lambda_1 \geq \lambda_2 \geq \cdots \geq \lambda_p$, the corresponding eigenvectors are $\boldsymbol{u}_1, \boldsymbol{u}_2, \ldots, \boldsymbol{u}_p$.

4. The variance contribution rate η_i and the accumulated variance contribution rate P are calculated by

$$\eta_i = \lambda_i / \sum_{i=1}^{p} \lambda_i \tag{3.25}$$

$$P = \sum_{i=1}^{k} \lambda_i / \sum_{i=1}^{p} \lambda_i \tag{3.26}$$

where λ_i is the eigenvalue of ith principal component, k is the number of the principal components, p is the total number of the original features, $k \leq p$. Generally, the former k principal components corresponding to the eigenvalues whose P is greater than 85–95% is selected to replace the original features, and therefore the number of the principal components is p, that is, the feature dimension is p-dimension.

3.5 Sample Selection Method

The selection of training samples has important influence on the classification performance and prediction results of the SVM model. If the SVM model can be trained by the least training samples and be applicable to accurately predict the discharge voltages of test sample set, then the test work and experimental cost can be greatly reduced. Hence, in order to improve the learning and generalization performance of the SVM model, the training sample set should be selected reasonably.

Given enough sample data, the training samples can be selected by orthogonal design method, with uniform dispersion and symmetrical comparability, so as to guarantee the generalization performance of the SVM model with least training samples. If the sample data is limited and cannot meet the level design requirements of the orthogonal array, the influence factors involved in the predicted object should be given consideration in training sample selection, thus to obtain good predicted results of air gap discharge voltages, with different gap structures and under different voltage waveforms.

Air gaps with typical electrodes, like the sphere gap, rod-plane and rod-rod gap are the pointcut to study air gap discharge characteristics, whose electric field distributions respectively represent the slightly uneven electric field, asymmetric severe nonuniform electric field and symmetric severe nonuniform electric field. Their discharge voltages are relatively easy to be measured by experiments. For discharge voltage prediction of an air gap, one or several kinds of typical gaps, whose electric field distributions are similar with the predicted air gap, can be selected as training samples to train the SVM model. Theoretically speaking, it should have better predicted results by this way. In order to predict the effects of voltage waveform (such as the front time) on air gap discharge voltage, the training samples should include the discharge voltage test data under different voltage waveforms, so as to establish the grey correlation between the voltage waveform features and the discharge voltage.

3.6 Error Analysis Method

In order to evaluate the predicted results of the SVM model, 4 error indexes including the sum of squared errors e_{SSE}, the mean square error e_{MSE}, the mean absolute percentage error e_{MAPE} and the mean square percentage error e_{MSPE} are applied for error analysis. Their calculation formulas are as follows.

$$e_{\mathrm{SSE}} = \sum_{i=1}^{N} [U_{\mathrm{t}}(i) - U_{\mathrm{p}}(i)]^2 \tag{3.27}$$

$$e_{\mathrm{MSE}} = \frac{1}{N}\sqrt{\sum_{i=1}^{N} [U_{\mathrm{t}}(i) - U_{\mathrm{p}}(i)]^2} \tag{3.28}$$

$$e_{\mathrm{MAPE}} = \frac{1}{N}\sum_{i=1}^{N} \left| \frac{U_{\mathrm{t}}(i) - U_{\mathrm{p}}(i)}{U_{\mathrm{t}}(i)} \right| \tag{3.29}$$

$$e_{\mathrm{MSPE}} = \frac{1}{N}\sqrt{\sum_{i=1}^{N} \left[\frac{U_{\mathrm{t}}(i) - U_{\mathrm{p}}(i)}{U_{\mathrm{t}}(i)} \right]^2} \tag{3.30}$$

where $U_{\mathrm{t}}(i)$ and $U_{\mathrm{p}}(i)$ are respectively the test value and the predicted result of the ith test sample, and N is the number of the test samples.

3.7 Implementation Process of the Prediction Model

The implementation process of the air gap discharge voltage prediction model based on SVM is shown in Fig. 3.7. The process is introduced in detail as follows.

1. The SVM model is trained by the training sample set $\{\Omega_1, \Omega_2, \ldots, \Omega_m\}$. Taking a training sample Ω_1 for example, firstly, the voltage intervals are classified. According to the basic ideas of the prediction method introduced in Sect. 1.2.2, set $a = 10\%$, the applied voltage is selected by step size of $0.01U_{\mathrm{b}}$ in the withstanding interval $[0.9U_{\mathrm{b}}, U_{\mathrm{b}})$ and breakdown interval $[U_{\mathrm{b}}, 1.1U_{\mathrm{b}}]$, therefore the voltage values in the withstanding interval are $\{0.9U_{\mathrm{b}}, 0.91U_{\mathrm{b}}, \ldots, 0.99U_{\mathrm{b}}\}$ and those in the breakdown interval are $\{U_{\mathrm{b}}, 1.01U_{\mathrm{b}}, \ldots, 1.1U_{\mathrm{b}}\}$, altogether 21 voltage values. By this way, the training sample data can be extended from 1 to 21. If the

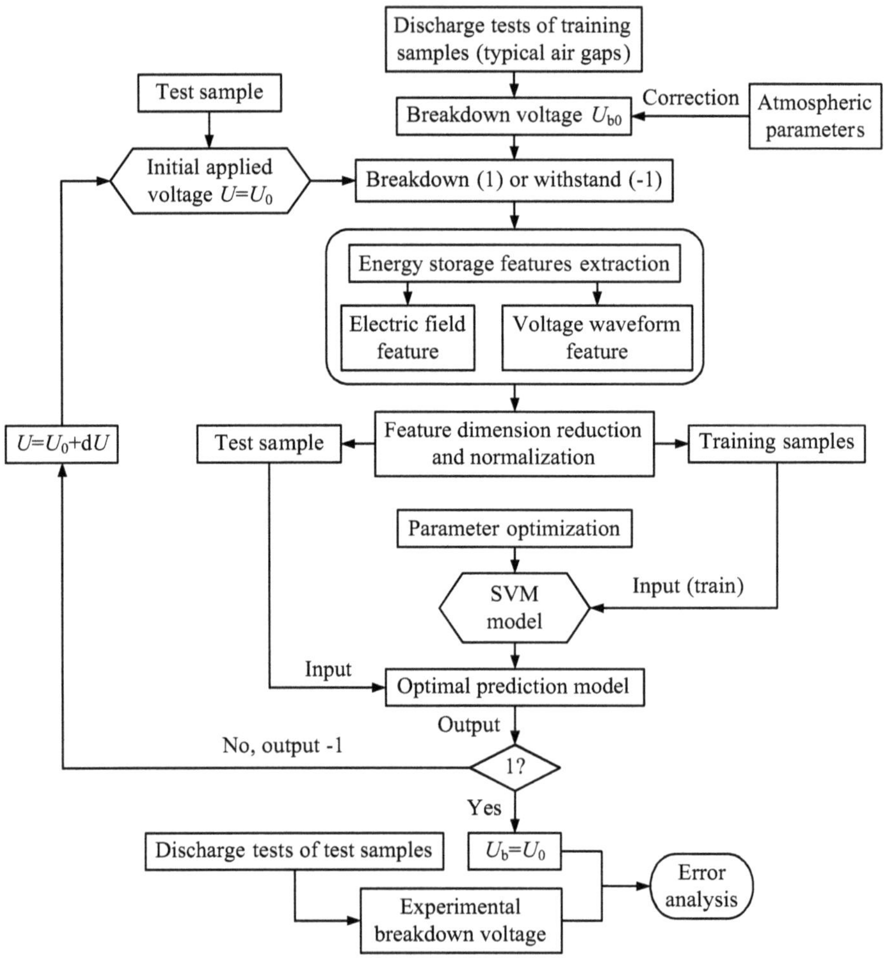

Fig. 3.7 Implementation process of the air gap discharge voltage prediction model

training sample set contains m experimental data of air gap discharge voltages, the number of samples actually used to train the SVM model is $21 \times m$.

Then, the simulation model of each air gap in the training sample set is established by a finite element analysis software to calculate the electric field distribution, while the applied voltage values are respectively those in the withstanding voltage interval and breakdown voltage interval. The electric field feature set $\boldsymbol{E}_1$ under each applied voltage is obtained by post processing of the electric field calculation results. In addition, the voltage waveform feature set $\boldsymbol{U}_1$ can be calculated according to their definitions and formulas.

Finally, the electric field features and the impulse voltage waveform features are summarized and normalized to the interval [0, 1], so as to eliminate the influence of different orders of magnitude and dimensions on the prediction results. So the energy storage feature set $\boldsymbol{F}_1$ of the training sample Ω_1 can be obtained and these features will be input to the SVM model. The outputs of the 10 voltage values in the withstanding interval are -1, while the 11 voltage values in the breakdown interval are 1. Using parameter optimization method to search the optimal parameter group (C, γ), under which the SVM model has the highest classification accuracy for the training sample set. Since the electric field features are not completely independent of each other, there may be some features with strong cross correlation and some features that are weakly correlated with the discharge voltage. These features may cause information redundancy and affect the classification performance of the SVM model. Hence, the energy storage feature dimension can be reduced before they are input to the SVM model.

The grey correlation between energy storage feature set and air gap discharge voltage can be directly established by the above method. The SVM model can be trained by the multi-dimensional nonlinear relations between the energy storage feature set $\boldsymbol{F}$ of the training sample set $\{\Omega_1, \Omega_2, \ldots, \Omega_m\}$ and the corresponding outputs -1 and 1. The process of extracting energy storage feature set and training SVM model is shown in Fig. 3.8.

2. The discharge voltages of the test sample set $\{\Psi_1, \Psi_2, \ldots, \Psi_n\}$ are predicted by the trained and optimized SVM model. Taking a test sample Ψ_1 for example, firstly, the estimated initial value of the discharge voltage $U = U_0$ is applied to extract the electric field features and the voltage waveform features. Then these features are summarized and normalized to construct the energy storage feature set, and they will be input (or after feature dimension reduction) to the SVM model for discharge voltage prediction.

If the 21 output values contain both -1 and 1, for example, the 1st, 2nd, …, $(p-1)$th output values are -1, and the pth, $(p+1)$th, …, 21st output values are 1, it means that when the applied voltage increases from $(0.9 + 0.01 \times (p-1))U$ to $(0.9 + 0.01 \times p)U$, the output will change from -1 to 1. Therefore the withstanding voltage interval is $[0.9U, (0.9 + 0.01 \times (p-1))U)$ and the breakdown voltage interval is $[(0.9 + 0.01 \times p)U, 1.1U]$. The voltage $(0.9 + 0.01 \times p)U$ is taken as the predicted result of the breakdown voltage.

If all of the 21 output values under the initial voltage $U = U_0$ are -1, then the applied voltage should be increased to $U = U_0 + \mathrm{d}U$ until the model outputs 1. On the contrary, if all of the 21 output values are 1, then the applied voltage should be decreased to $U = U_0 - \mathrm{d}U$ until the model outputs -1. This method is similar to the up-and-down method applied in the impulse discharge test. $\mathrm{d}U$ can be determined according to experience and accuracy requirement. Hence, when the output of the SVM model changes from -1 to 1, the applied voltage corresponding to the lower limit of the voltage interval whose outputs are 1 is taken as the discharge voltage prediction result U_p. Finally, U_p is compared with the experimental value U_t for error analysis. The flow chart of discharge voltage prediction by SVM model is shown in Fig. 3.9.

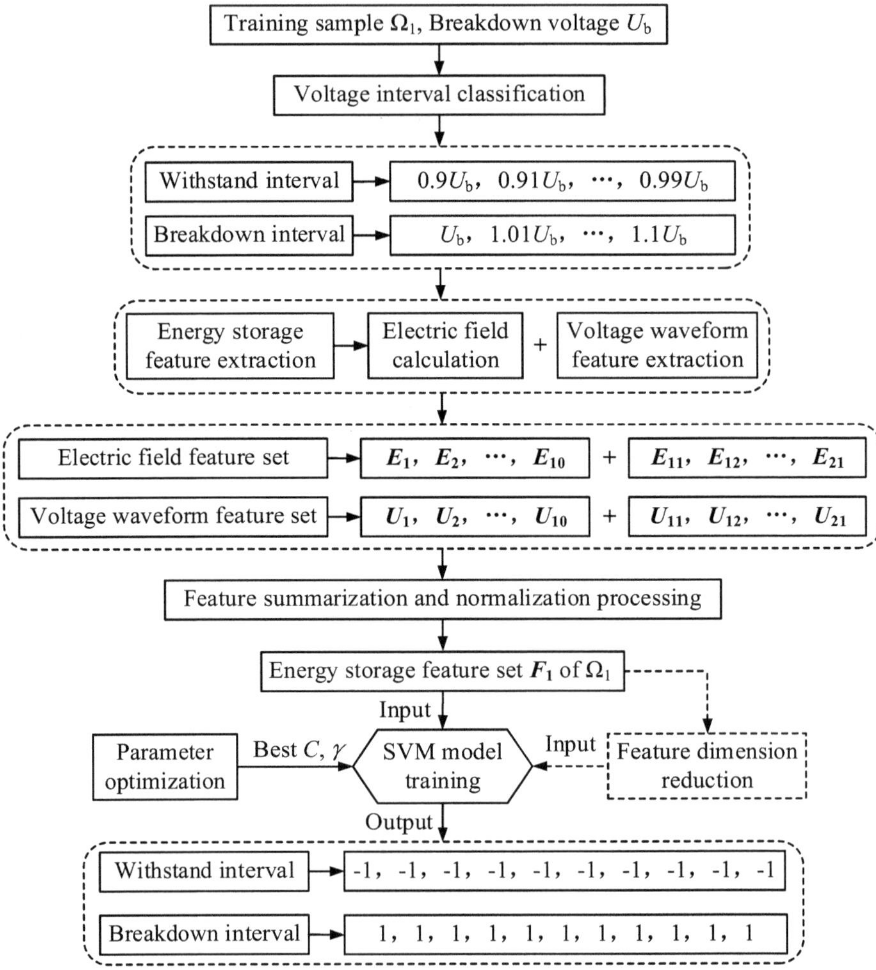

Fig. 3.8 Flow chart of extracting energy storage feature set and training SVM model

3.8 Brief Summary

In this chapter, the air gap discharge voltage prediction model based on SVM is introduced. The relevant theoretical basis for the realization of this method is presented, including the fundamental theory of SVM, three kinds of parameter optimization methods, namely, the improved GS algorithm, GA and PSO algorithm. The feature dimension reduction methods, the sample selection method and the error analysis method are also introduced. Finally, the procedure to achieve air gap discharge voltage prediction by SVM is described in detail.

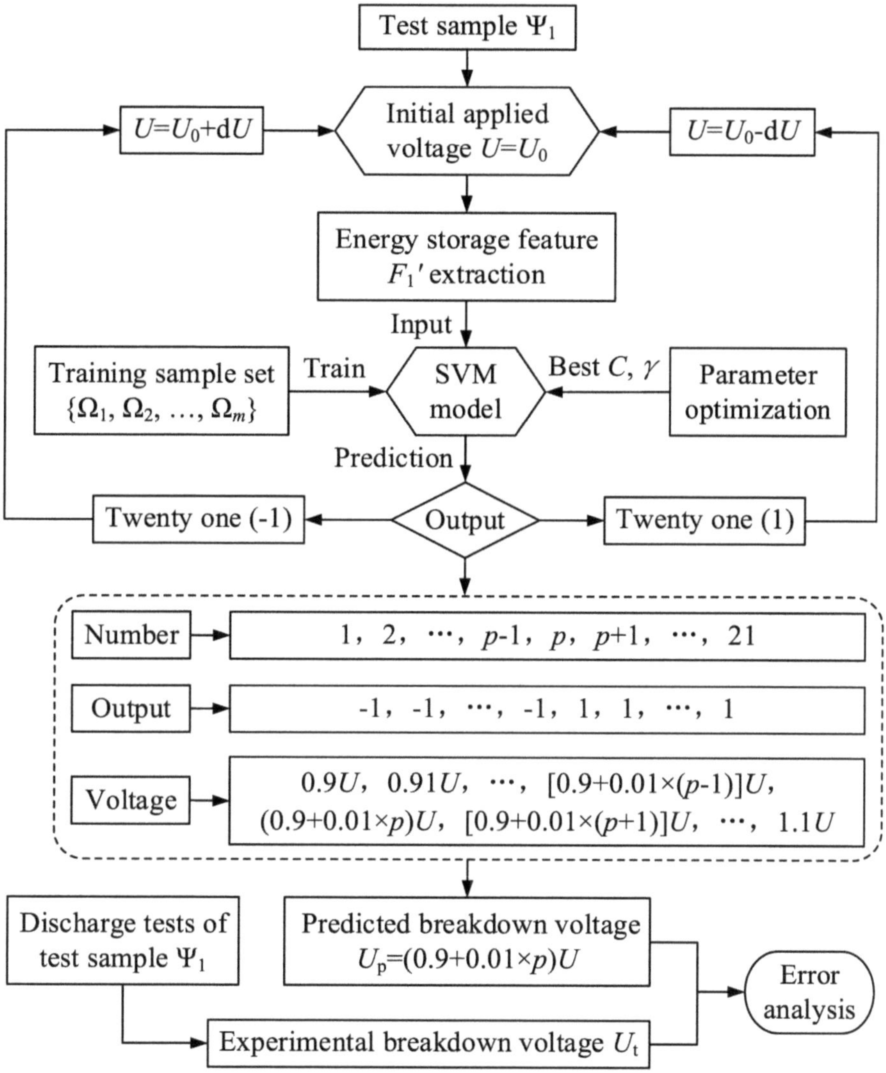

Fig. 3.9 Flow chart of discharge voltage prediction by SVM model

References

1. Bourek Y, Mokhnache L, Said NN et al (2009) Study of discharge in point-plane air interval using fuzzy logic. J Electr Eng Technol 4(3):410–417
2. Bourek Y, Mokhnache L, Said NN et al (2011) Determination of ionization conditions characterizing the breakdown threshold of a point-plane air interval using fuzzy logic. Electr Power Syst Res 81(11):2038–2047

3. Ruiz D, Llovera-Segovia P, Pomar V et al (2013) Analysis of breakdown process at U50 voltage for plane rod discharges by means of neural networks. J Electrostat 71(3):336–340
4. Mokhnache L, Boubakeur A (2001) Prediction of the breakdown voltage in a point-barrier-plane air gap using neural networks. In: 2001 annual report conference on electrical insulation and dielectric phenomena, Kitchener, Ontario, Canada, 14–17 October 2001
5. Mokhnache L, Boubakeur A, Feliachi A (2004) Breakdown voltage prediction in a point-barrier-plane air gap arrangement using self-organization neural networks. In: IEEE power engineering society general meeting, Denver, Colorado, USA, 6–10 June 2004
6. Hu Q, Shu LC, Jiang XL et al (2010) Conductor's ac corona onset voltage prediction under different atmospheric parameters and conductor surface conditions. High Volt Eng 36(7):1669–1674
7. Yuan Y, Jiang XL, Du Y et al (2012) Predictions of the AC discharge voltage of short rod-plane air gap under rain conditions with the application of ANN. High Volt Eng 38(1):102–108
8. Zhang GB, Luo X, Shen YY et al (2014) Effect of atmosphere condition on discharge characteristics of air gap and the application of neural network. High Volt Eng 40(2):564–571
9. Luo X, Niu HQ, Lin HR et al (2013) Application and contrast analysis of BP and RBF neural network in prediction of breakdown voltage of air gap. Adv Technol Electr Eng Energy 32(3):110–115
10. Niu HQ, Xu J, Wu JZ et al (2017) Gray correlation analysis of atmospheric conditions and prediction of air gap discharge voltage. J South China Univ Technol (Nat Sci Ed) 45(7):48–54
11. Jiang H, Zhang B, Lian XX et al (2014) Research on prediction method of I-air breakdown voltage based on ANN. Power Syst Clean Energy 30(9):5–11
12. Shu SW (2014) Study on prediction for corona onset and breakdown voltages of air gap based on electric field features and support vector machine. Dissertation, Wuhan University, Wuhan
13. Vapnik VN (1995) The nature of statistical learning theory. Springer, New York
14. Deng NY, Tian YJ (2004) New methods in data mining—support vector machine. Science Press, Beijing
15. Deng NY, Tian YJ (2009) Support vector machine—theory, algorithm and development. Science Press, Beijing
16. Chang TT (2010) Research on some problems of support vector machine learning algorithm. Dissertation, Xidian University, Xi'an
17. Stone M (1974) Cross-validatory choice and assessment of statistical predictions. J R Stat Soc Ser B (Methodol) 36(2):111–147
18. MATLAB Chinese Forum (2010) Thirty cases analysis of MATLAB neural network. Beijing University of Aeronautics and Astronautics Press, Beijing
19. Wang JF (2012) Study on speaker recognition based on improved grid search parameters optimization algorithm of SVM. Dissertation, Harbin Engineering University, Harbin
20. Yang SY (2011) Pattern recognition and intelligent computing—Matlab technical realization, 2nd edn. Publishing House of Electronics Industry, Beijing

Chapter 4
Corona Onset Voltage Prediction of Electrode Structures

4.1 Corona Discharge

Corona effect is the acoustic, optical and thermal phenomena appeared during the process of gas ionization and recombination in the corona discharge. The corona of high voltage transmission line will cause a series of consequences, including corona loss, ion flow, radio interference and audible noise. Corona effect has become one of the decisive factors in the conductor selection of UHV transmission lines. The surface electric field intensities of the fittings in high voltage transmission and transformation projects, such as grading rings and spacers, are usually very high due to the small curvature radius. Once the electric field strength exceeds the critical value, corona discharge happens, and the resulted electromagnetic environment problems have gotten more and more attention. In addition, the corona discharge will produce ozone and nitrogen oxides, which will react with the surrounding medium, and results in accelerated aging of line insulation and corrosion of metal components. In order to evaluate the corona effects of transmission lines and fittings, it is necessary to understand the basic characteristics of corona discharge, and obtain the corona onset voltage and electric field strength, so as to control the corona effect within a reasonable range.

4.1.1 Basic Characteristics of Corona Discharge

Before complete breakdown of an air gap in an extremely nonuniform electric field, a thin luminous layer will generate near the electrode with large curvature, this phenomenon is called corona discharge. Corona discharge has obvious polarity effect. Taking rod plane gap for example, if the rod electrode is with negative polarity, when the voltage rises to a certain value and the average current is close to microampere level, a repetitive and regular pulse current will appear. When the voltage rises, the amplitude of the current pulse is almost invariant, but the frequency will increase

Z. Qiu et al., *Air Insulation Prediction Theory and Applications*, Power Systems,
https://doi.org/10.1007/978-981-10-5163-0_4

and the average current increases accordingly. When the voltage continues to rise, the corona current loses the property of regular high frequency pulse, and converts to sustained current, while the average current still increases with the rising voltage. When the voltage rises further, the irregular current pulses with much larger amplitude (streamer corona current) will appear and develop into brush discharge. The brush discharge is a kind of partial discharge more intense than the corona discharge. The air gap will breakdown quickly when the voltage increases after the brush discharge appears.

If the rod electrode is with positive polarity, the corona current also has the property of repetitive pulses, but with an irregular pattern. When the voltage continues to rise, the pulse characteristics of the current become less and less significant until the pulse current turns into sustained current. When the voltage rises further, the irregular streamer corona current pulses with much larger amplitudes will appear, that is the brush discharge.

The corona discharge in air has the following effects [1, 2]:

1. It has acoustic, optical and thermal effects. The corona discharge will produce hissing sound and emit blue purple halo, which will heat the surrounding gas and cause energy loss.
2. Corona will produce high frequency pulse current containing many higher harmonics, which will cause radio interference. With the increase of transmission line voltage class, the radio interference caused by line corona has become one of the key concerns in the design and operation of transmission lines.
3. Corona generates audible noise, and it will cause physiological and psychological effects on people. The problem of audible noise is not serious for the power transmission and transformation systems under 500 kV, but for UHV transmission lines, how to reduce the surface field intensity of the conductor and the audible noise is a key issue.
4. At the tip or some protrusions of the electrode, the electrons and ions will move with high speed under the local strong electric field and they will exchange momentum with the gas molecules, which will produce "electrical wind" and cause vibrations of the electrode or the conductor.
5. Corona discharge will generate some chemical reactions, and produce ozone (O_3), nitric oxide (NO) and nitrogen dioxide (NO_2) in the air. O_3 is a strong oxidant that has strong oxidation effects on metals and organic insulation. NO and NO_2 can react with the moisture in air and produce highly corrosive nitric acid.

In addition, the corona discharge also has some advantages. Corona loss can reduce the amplitude and steepness of the lightning impulse voltage wave on transmission lines and attenuate the switching overvoltage. The corona discharge can also improve the electric field distribution. Moreover, the corona effect can be applied to make electrostatic precipitator, ozone generator and electrostatic spraying equipment.

4.1.2 Corona Onset Voltage and Inception Field Strength

The voltage making corona inception is called corona onset voltage U_c, and the field strength on the electrode surface is called corona inception field strength E_c. The corona onset voltage is lower than the breakdown voltage, and their difference is more significant in more inhomogeneous electric field. Theoretically speaking, the corona onset voltage can be calculated according to the self-sustained discharge condition $\gamma \exp \int_0^d (\alpha - \eta)\mathrm{d}x = 1$, where d is the thickness of the corona layer, determined by the condition of $\alpha = \eta$ at the edge of the corona layer. Since the calculation is complicated, and the theoretical calculation itself is not accurate, the corona onset voltage is actually estimated by empirical formulas summarized by experimental results [3].

For a single conductor with the radius of r and the height over the ground is h, the relation between the conductor surface field strength E and the voltage to ground U is

$$E = \frac{U}{r \ln \frac{2h}{r}} \tag{4.1}$$

For parallel conductors, when the distance d between the axes is much greater than the conductor radius r, the relation between the conductor surface field strength E and the voltage to the neutral plane U (half of the voltage between conductors) is

$$E = \frac{U}{r \ln \frac{d}{r}} \tag{4.2}$$

Therefore, the corona inception field strength E_c can be calculated after measuring the corona onset voltage U_c. F. W. Peek conducted numerous experiments, and the results showed that the E_c is related to the electrode size and the climate condition. According to the American Standard, taking 760 mmHg and 25 °C as the standard condition, Peek proposed the empirical formulas for corona inception field strength (peak value) of parallel conductors, coaxial cylinder and sphere-sphere electrode structures, respectively expressed as [4]:

$$E_c = 29.8\delta\left(1 + \frac{0.301}{\sqrt{r\delta}}\right)(\mathrm{kV/cm}) \tag{4.3}$$

$$E_c = 31\delta\left(1 + \frac{0.308}{\sqrt{r\delta}}\right)(\mathrm{kV/cm}) \tag{4.4}$$

$$E_c = 27.2\delta\left(1 + \frac{0.34}{\sqrt{r\delta}}\right)(\mathrm{kV/cm}) \tag{4.5}$$

where δ is the relative air density.

According to IEC standards, taking 101.3 kPa and 20 °C as standard condition, the Eqs. 4.3–4.5 can be modified to [3]:

$$E_c = 30.3\delta\left(1 + \frac{0.298}{\sqrt{r\delta}}\right)(\text{kV/cm}) \quad (4.6)$$

$$E_c = 31.5\delta\left(1 + \frac{0.305}{\sqrt{r\delta}}\right)(\text{kV/cm}) \quad (4.7)$$

$$E_c = 27.7\delta\left(1 + \frac{0.337}{\sqrt{r\delta}}\right)(\text{kV/cm}) \quad (4.8)$$

The surface condition of the conductor has a great influence on the corona inception field strength E_c. For stranded conductor, the calculation formula of E_c needs to be modified by multiplying the surface roughness coefficient m on the basis of Eq. 4.6, i.e.,

$$E_c = 30.3m\delta\left(1 + \frac{0.298}{\sqrt{r\delta}}\right)(\text{kV/cm}) \quad (4.9)$$

For ideal smooth cylindrical conductor, $m = 1$, and for clean stranded conductor, $m = 0.75$–0.85. Presence of scratches on the conductor surface may reduce m to 0.6–0.8. Extreme conditions such as insects, plants, water, ice, snow, etc. may further reduce m to 0.3–0.6 [5].

4.2 Corona Onset Voltage Prediction of Rod-Plane Electrodes

The proposed air insulation prediction model is used to predict the corona onset voltages of hemispherical rod-plane electrodes with different rod diameters. The predicted results are compared with the calculation results of the existing methods, including the empirical and semi-empirical formulas, critical charge criterion, and the photoionization model.

4.2.1 Training and Test Sample Set

The structure diagram of a rod-plane gap with a hemispherical rod head is shown in Fig. 4.1. The rod radius is R and the gap distance is d. A two-dimensional axisymmetric model on xz plane is established to calculate electric field distribution.

The experimental data of the positive DC corona onset voltages of rod-plane electrodes were collected from [6–8], as shown in Table 4.1, where R is the rod radius, d is the gap distance, and U_c is the corona onset voltage. 8 data are selected from Table 4.1 as training samples, and the other 36 data are test samples.

The electric field distribution of the rod-plane is calculated. The rod electrode is applied 1 V, and the plane electrode is applied zero potential. Figure 4.2a, b

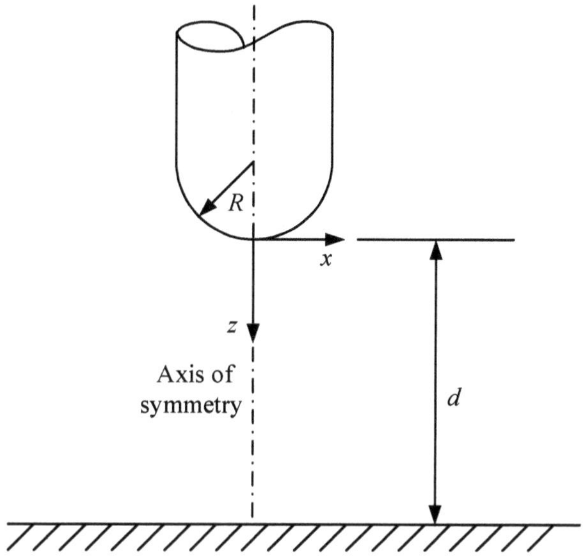

Fig. 4.1 Structure diagram of the rod-plane gap

Table 4.1 Positive DC corona onset voltages of rod-plane electrodes: (**a**) training samples, (**b**) experimental data of test samples

R (mm)	d (cm)	U_c (kV)	R (mm)	d (cm)	U_c (kV)
a					
2.0	1.0	13.89	2.0	5.0	19.17
	2.0	16.13	4.0	4.0	27.29
	3.0	17.23		5.0	28.72
	4.0	18.33	10.0	5.0	47.00
b					
1.5	1.0	11.28	4.0	1.0	18.14
	1.5	12.04		1.5	20.28
	2.0	12.81		2.0	22.57
	3.0	14.33		2.5	23.63
	4.0	15.40		3.0	24.71
2.0	1.5	15.24		3.5	26.00
	2.5	17.02		4.5	28.00
	3.5	17.96		5.5	29.40
	4.5	18.61		6.0	29.89
	5.5	19.93	10.0	1.0	25.67
	6.0	20.11		2.0	35.53
	6.5	20.48		3.0	41.00
	7.0	20.66		4.0	44.44
	7.5	20.85		6.0	48.44
	8.0	20.94		10.0	54.06

(continued)

Table 4.1 (continued)

R (mm)	d (cm)	U_c (kV)	R (mm)	d (cm)	U_c (kV)
	8.5	21.20		15.0	57.82
	9.0	21.39		20.0	60.39
	9.5	21.76		25.0	62.42

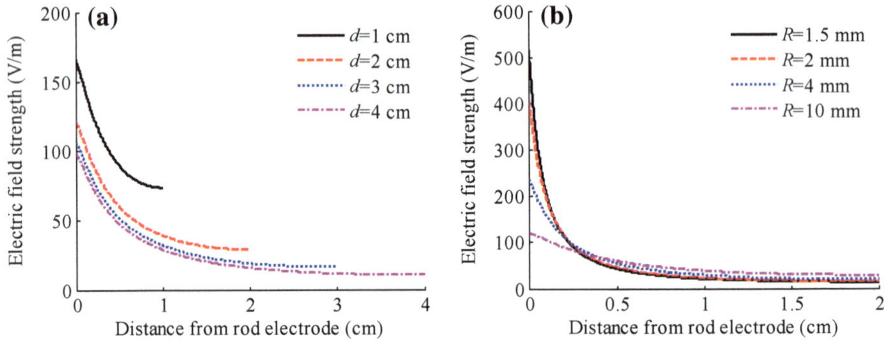

Fig. 4.2 Electric field distributions on the shortest path of rod-plane gaps: **a** R = 10 mm, d = 1–4 cm, **b** d = 2 cm, R = 1–10 mm

respectively show the electric field distributions on the shortest path of the rod-plane gap when the gap distance ranging from 1 cm to 4 cm (R = 10 mm) and the rod radius ranging from 1 mm to 10 mm (d = 2 cm).

It can be seen from Fig. 4.2 that the maximum electric field strength on the electrode surface decreases with the increased distance under the invariant radius. But when the distance increases to a certain value, the decrease of the maximum field strength is not obvious. The reduction rate of the field strength on the shortest path decreases with the increase of the gap distance. In addition, the maximum field strength of the electrode surface decreases obviously with the increase of the rod radius under the invariant gap distance, and the reduction rate of the field strength on the shortest path decreases with the increase of the rod radius.

For the training sample set in Table 4.1, the corona onset voltage is applied on the rod electrode and the plane electrode is applied zero potential to calculate the electric field distribution. The calculation results are post-processed to extract the original data, such as the electric field values on the nodes, elements or along a path, the area and volume of elements, and the coordinates of the nodes or the element barycenter. According to the definitions and calculation formulas introduced in Sect. 2.2.1, the electric field features of the gap under the corona onset voltage can be calculated. For corona onset voltage prediction, the electric field features including those belong to the discharge channel, the electrode surface and the shortest path, as shown in Table 2.1, while the 8 features related to 24 and 7 kV/cm are rejected. All of the electric field features are 29 dimensions.

4.2.2 SVM Prediction Results and Analysis

The improved GS algorithm is applied for parameter optimization of the SVM model under the sense of 3-CV, and the correlation analysis method is used to reduce the electric field feature dimensions, so as to analyze the influence of different feature dimensions on the prediction results. The SVM prediction results of the positive DC corona onset voltages of rod-plane electrodes under different feature dimensions are shown in Table 4.2, where U_{t} and U_{p} are respectively the test value and prediction value of the corona onset voltage, and σ is the relative error. Thereinto, 29-dimension (29-d) is all of the electric field features, 25-d and 18-d are reduced from 29-d respectively by rejecting the features whose correlation coefficient with the corona onset voltage is less than 0.1 and 0.3, and 13-d and 7-d are respectively reduced from 25-d and 18-d by rejecting the features whose cross correlation coefficient is greater than 0.9.

The optimal parameters and errors of the prediction model under different feature dimensions are shown in Table 4.3. It can be seen from that the prediction accuracy varies with the feature dimensions. The prediction accuracy of 29-d is the highest and the e_{MAPE} is only 1.72%. The prediction errors of 25-d and 18-d are larger than those of 29-d, which indicates that the rejected electric field features contain effective information about the non-linear relationship with the corona onset voltage. In the condition of 13-d or 7-d, the eMAPE and eMSPE increase, but the eSSE and eMSE decrease compared with 25-d and 18-d. In general, the effect on the prediction accuracy is not significant by rejecting the electric field features whose cross correlation coefficient is greater than 0.9.

4.2.3 Comparison with Other Prediction Methods

4.2.3.1 Critical Charge Criterion

According to the transition mechanism of electron avalanche into streamer, when the charge number in the head of the initial electron avalanche reaches a critical value, the secondary electron avalanches will develop and transform to streamer due to the strong enough electric field distortion. As shown in Fig. 4.3, the initial electron avalanche starts from $z = z_i$, where the collision ionization coefficient α is equal to the electron attachment coefficient η. In the ionized area, $\alpha > \eta$. At the boundary of the corona layer, $\alpha = \eta$, the collision ionization does not occur, and the free electrons stop growth. Only when the charge number in the head of the initial electron avalanche reaches a certain value, the electric field will be distorted and strengthened to a certain extent until the space photoionization effect is strong enough, then the electron avalanche will transform to streamer.

The critical charge criterion thinks that the streamer incepts when the charge in the head of the electron avalanche satisfies the following relation:

Table 4.2 Positive DC corona onset voltage prediction results of rod-plane electrodes under different feature dimensions: (**a**) $R = 1.5$ mm, (**b**) $R = 2$ mm, (**c**) $R = 4$ mm, (**d**) $R = 10$ mm

d (cm)	U_t (kV)	29-d		25-d		18-d		13-d		7-d	
		U_p (kV)	σ (%)	U_p (kV)	σ (%)	U_p (kV)	σ (%)	U_p (kV)	σ (%)	U_p (kV)	σ (%)
a											
1.0	11.28	11.39	1	11.51	2	11.62	3	11.96	6	11.84	5
1.5	12.04	12.28	2	12.52	4	15.52	4	12.76	6	12.64	5
2.0	12.81	12.94	1	13.19	3	13.19	3	13.45	5	13.19	3
3.0	14.33	13.90	−3	14.04	−2	14.19	−1	14.47	1	14.04	−2
4.0	15.40	14.63	−5	14.78	−4	14.94	−3	15.09	−2	14.78	−4
b											
1.5	15.24	15.09	−1	15.09	−1	15.09	−1	15.09	−1	15.24	0
2.5	17.02	16.85	−1	16.68	−2	16.68	−2	17.02	0	16.68	−2
3.5	17.96	17.78	−1	17.78	−1	17.78	−1	17.96	0	17.78	−1
4.5	18.61	18.61	0	18.61	0	18.80	1	18.61	0	18.61	0
5.5	19.93	19.33	−3	19.33	−3	19.53	−2	19.53	−2	19.53	−2
6.0	20.11	19.91	−1	19.71	−2	19.91	−1	19.91	−1	19.91	−1
6.5	20.48	20.07	−2	20.07	−2	20.28	−1	20.28	−1	20.28	−1
7.0	20.66	20.45	−1	20.45	−1	20.66	0	20.66	0	20.66	0
7.5	20.85	20.64	−1	20.64	−1	20.85	0	20.85	0	21.06	1
8.0	20.94	20.94	0	20.94	0	21.36	2	21.15	1	21.36	2

(continued)

Table 4.2 (continued)

d (cm)	U_t (kV)	29-d		25-d		18-d		13-d		7-d	
		U_p (kV)	σ (%)	U_p (kV)	σ (%)	U_p (kV)	σ (%)	U_p (kV)	σ (%)	U_p (kV)	σ (%)
8.5	21.20	21.41	1	21.20	0	21.62	2	21.41	1	21.84	3
9.0	21.39	21.60	1	21.39	0	22.03	3	21.82	2	22.03	3
9.5	21.76	21.98	1	21.76	0	22.20	2	22.20	2	22.63	4
c											
1.0	18.14	18.50	2	19.77	2	17.23	−5	19.41	7	16.87	−7
1.5	20.28	21.29	5	22.11	5	21.29	5	21.50	6	21.09	4
2.0	22.57	23.02	2	23.70	5	23.92	6	23.25	3	23.47	4
2.5	23.63	24.34	3	24.81	5	25.05	6	24.58	4	25.05	6
3.0	24.71	25.70	4	25.95	5	25.95	5	25.45	3	25.95	5
3.5	26.00	26.52	2	26.78	3	26.78	3	26.52	2	26.78	3
4.5	28.00	28.00	0	28.00	0	28.00	0	28.00	0	28.28	1
5.5	29.40	29.40	0	29.11	−1	29.11	−1	29.11	−1	29.11	−1
6.0	29.89	29.89	0	29.59	−1	29.59	−1	29.89	0	29.59	−1
d											
1.0	25.67	24.39	−5	23.10	−5	24.39	−5	24.64	−4	23.10	−10
2.0	35.53	34.11	−4	35.89	2	37.31	5	34.46	−3	31.98	−10
3.0	41.00	40.18	−2	41.82	1	43.46	6	39.77	−3	39.36	−4
4.0	44.44	44.44	0	43.55	2	46.22	4	44.00	−1	44.44	0

(continued)

Table 4.2 (continued)

d (cm)	U_t (kV)	29-d		25-d		18-d		13-d		7-d	
		U_p (kV)	σ (%)	U_p (kV)	σ (%)	U_p (kV)	σ (%)	U_p (kV)	σ (%)	U_p (kV)	σ (%)
6.0	48.44	47.96	−1	46.99	−2	46.99	−3	49.89	3	48.44	0
10.0	54.06	53.52	−1	50.82	−3	50.28	−7	56.76	5	53.52	−1
15.0	57.82	56.66	−2	54.35	−6	55.51	−4	60.13	4	56.66	−2
20.0	60.39	59.79	−1	56.77	−4	57.97	−4	62.81	4	59.18	−2
25.0	62.42	61.17	−2	57.43	−4	59.92	−4	64.92	4	61.17	−2

Table 4.3 Optimal parameters and errors of SVM prediction model under different feature dimensions

Parameters and errors	29-d	25-d	18-d	13-d	7-d
C	315.173	137.187	274.374	362.039	128
γ	0.2333	0.25	0.25	0.25	0.1436
e_{SSE}	12.9796	38.7602	58.0021	38.3195	37.5063
e_{MSE}	0.1001	0.1729	0.2116	0.1720	0.1701
e_{MAPE}	0.0172	0.0233	0.0294	0.0244	0.0283
e_{MSPE}	0.0037	0.0048	0.0059	0.0053	0.0063

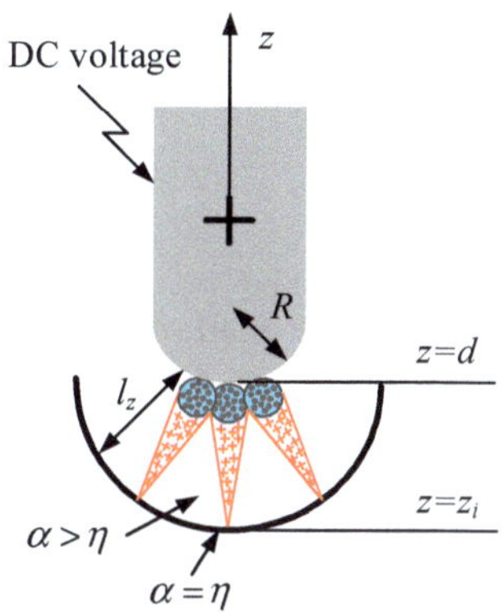

Fig. 4.3 Schematic diagram of the streamer inception process under DC voltage

$$\exp \int_{z=z_i}^{z=d} (\alpha - \eta)\mathrm{d}z \geq N_{\mathrm{crit}} \tag{4.10}$$

where d is the gap distance, N_{crit} is the critical charge number. α and η are both the function of E/p, where E is the electric field strength and p is the pressure. N_{crit} is usually taken as 10^8 [9].

Afterwards, some researchers found that the value of N_{crit} should be modified according to the electrode structure and size, in order to obtain the calculated results that are in consistent with the test values. For example, N_{crit} was determined 10^4 in [10] for rod-plane electrodes and conductors with the radius ranging from 0.01 to 20 cm. The hemispherical rod-plane electrodes shown in Table 4.1 are with the rod radius of 1.5, 2, 4 and 10 mm, which are in the range of application for $N_{\mathrm{crit}} = 10^4$. Substituting it into Eq. 4.10, it can be derived that

$$\int_{z=z_i}^{z=d} (\alpha - \eta)\mathrm{d}z \geq 9.21 \tag{4.11}$$

Fig. 4.4 Flow chart of corona onset voltage calculation based on the critical charge criterion

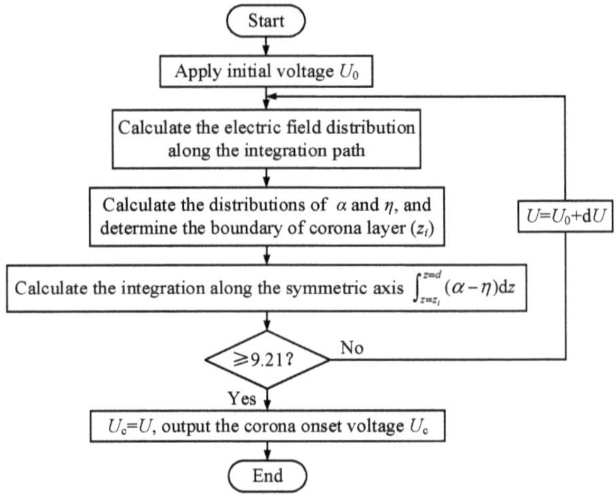

The flow chart of corona onset voltage calculation by the critical charge criterion is shown in Fig. 4.4 [11]. Taking dU as the step size, the applied voltage is increased continuously and the spatial electric field distribution is calculated by FEM. The functional relationships between α, η and the electric field strength are substituted into Eq. 4.11. When the equal sign holds up, the corresponding applied voltage is the corona onset voltage. Using the critical charge criterion to calculate the corona onset voltages of the test samples shown in Table 4.1, the results are shown in Table 4.4.

The error indexes of the calculation results by critical charge criterion are $e_{\text{SSE}} = 203.7565$, $e_{\text{MSE}} = 0.3965$, $e_{\text{MAPE}} = 0.0679$, $e_{\text{MSPE}} = 0.0143$. It can be seen that calculated results are with large errors, especially the absolute percentage errors of some test samples exceed 10%. The results indicate that the applicability of $N_{\text{crit}} = 10^4$ in this criterion is limited.

4.2.3.2 Photoionization Model

The space photoionization has an important effect on streamer inception and propagation. By considering the effect of photons on the formation of secondary electron avalanches, E. Nasser proposed another streamer inception condition, that is, the electrons produced in secondary electron avalanches are more than those generated in the initial electron avalanche [12]. This method is usually called the photoionization model. It has been widely used in the numerical simulation of streamer inception and the calculation of the corona onset voltage.

Taking the rod-plane gap for example, when the rod electrode is applied positive DC voltage, the electric field strength around the rod head will increase with the voltage. In the area where $\alpha > \eta$, the ionizing collision of the air molecules will take place during the movement process of free electrons towards the rod. The initial

Table 4.4 Positive DC corona onset voltage calculation results of rod-plane electrodes by critical charge criterion: (**a**) $R = 1.5$ mm, (**b**) $R = 2$ mm, (**c**) $R = 4$ mm, (**d**) $R = 10$ mm

d (cm)	Test value (kV)	Calculated value (kV)	Relative error (%)
a			
1.0	11.28	11.50	1.95
1.5	12.04	12.37	2.74
2.0	12.81	12.99	1.41
3.0	14.33	13.75	−4.05
4.0	15.40	14.23	−7.60
b			
1.5	15.24	14.65	−3.87
2.5	17.02	15.87	−6.76
3.5	17.96	16.66	−7.24
4.5	18.61	17.17	−7.74
5.5	19.93	17.39	−12.74
6.0	20.11	17.56	−12.68
6.5	20.48	17.69	−13.62
7.0	20.66	17.74	−14.13
7.5	20.85	17.78	−14.72
8.0	20.94	17.90	−14.52
8.5	21.20	17.95	−15.33
9.0	21.39	18.01	−15.80
9.5	21.76	18.30	−15.90
c			
1.0	18.14	18.80	3.64
1.5	20.28	21.22	4.64
2.0	22.57	22.73	0.71
2.5	23.63	23.85	0.93
3.0	24.71	24.67	−0.16
3.5	26.00	25.26	−2.85
4.5	28.00	26.21	−6.39
5.5	29.40	26.85	−8.67
6.0	29.89	27.22	−8.93
d			
1.0	25.67	26.09	1.64
2.0	35.53	36.05	1.46
3.0	41.00	41.20	0.49
4.0	44.44	44.44	0
6.0	48.44	48.25	−0.39
10.0	54.06	51.82	−4.14
15.0	57.82	53.88	−6.81
20.0	60.39	54.98	−8.96
25.0	62.42	55.68	−10.80

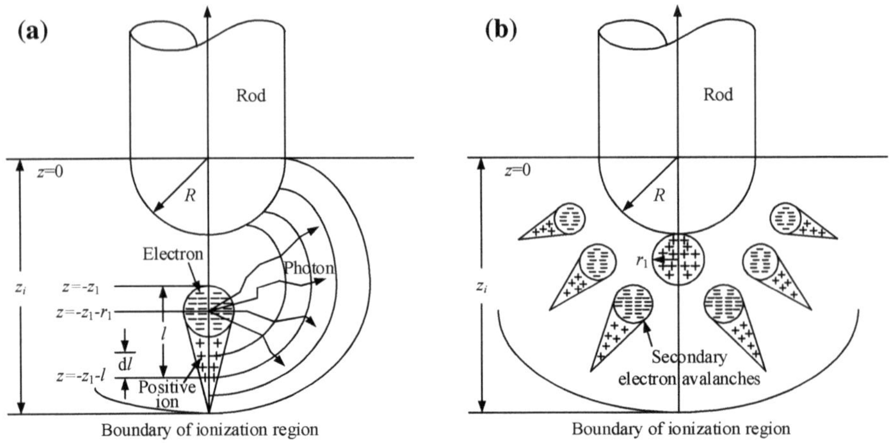

Fig. 4.5 Development process of initial electron avalanche and secondary electron avalanches of positive rod-plane gap: **a** initial electron avalanche, **b** secondary electron avalanche

electron avalanche will produce at the boundary of the ionization region ($z = -z_i$), as shown in Fig. 4.5a. At the position of $z = -z_1$, the electron number is:

$$N_1 = \exp\left[\int_{-z_i}^{-z_1} (\alpha - \eta)\mathrm{d}z\right] \tag{4.12}$$

The head of the electron avalanche may be assumed to have a spherical shape, its radius is:

$$r_1 = \sqrt{6\int_{-z_\mathrm{i}}^{-z_1} \frac{D_\mathrm{e}}{v_\mathrm{e}}\mathrm{d}z} \tag{4.13}$$

where D_e is the electron diffusion coefficient and v_e is the electron drift velocity, whose expressions are respectively:

$$D_\mathrm{e} = 5.3 \times 10^5 \left(\frac{E}{P}\right)^{1.21} / E \tag{4.14}$$

$$v_\mathrm{e} = \begin{cases} 1.0 \times 10^6 \times \left(\frac{E}{P}\right)^{0.715} & \frac{E}{P} \le 100 \\ 1.55 \times 10^6 \times \left(\frac{E}{P}\right)^{0.62} & \frac{E}{P} > 100 \end{cases} \tag{4.15}$$

The collision ionization of free electrons and air molecules will also make the molecules get excited and radiate photons to surroundings. The photoionization process will take place between these photons and air molecules, and the generated photoelectrons will flow to the rod electrode due to the effect of the total electric field. The secondary electron avalanches will be produced due to the collision ionization of photoelectrons with air molecules, as shown in Fig. 4.5b. If the number of photons radiated by one collision ionization is f_1, then the number of photons radiated by the electron avalanche head is $f_1 N_1$. The area from $z = -z_i$ to $z = -z_1 - r_1$ can be divided by the idea of numerical integration, while the thickness per layer is dl. At the position of $z = -z_1 - l$, the number of the electrons absorbed by this layer is $\mu f_1 N_1 \mathrm{e}^{-\mu l} g(l)$, where μ is the photon absorption coefficient and $g(l)$ is the geometrical factor used to consider the situation that some photons are not absorbed and disappear in the electrode. If the photoionization probability of air molecules by absorbing photons is f_2, then the number of the photoelectrons generated by photoionization in this layer is $\mu f_1 f_2 N_1 \mathrm{e}^{-\mu l} g(l)$, and the number of the secondary electron avalanches produced by collision ionization of air molecules is:

$$f_1 f_2 N_1 \mu e^{-\mu l} g(l) \exp\left[\int_{-(z_1+l)}^{-(z_1+r_1)} (\alpha - \eta)\mathrm{d}z\right] \tag{4.16}$$

The negative ions in the electron avalanche can be ignored, and therefore the total number of positive ions N_2 in secondary electron avalanches can be calculated by integral of the positive ion number generated in each layer. N_2 can be expressed as:

$$N_2 = \int_{-z_i}^{-(z_1+r_1)} f_1 f_2 N_1 \mu e^{-\mu l} g(l) \exp\left[\int_{-(z_1+l)}^{-(z_1+r_1)} (\alpha - \eta)\mathrm{d}z\right] \mathrm{d}l \tag{4.17}$$

Since most of the photons produced in the main electron avalanche are radiated during the last few steps of the electron avalanche growth, therefore, $|z_1| \approx R$. According to Nasser's criterion, the self-sustaining condition of positive DC corona of the rod-plane gap is:

$$\int_{-z_i}^{-(z_1+r_1)} f_1 f_2 N_1 \mu e^{-\mu l} g(l) \exp\left[\int_{-(z_1+l)}^{-(z_1+r_1)} (\alpha - \eta)\mathrm{d}z\right] \mathrm{d}l \geq N_1 \tag{4.18}$$

The positive DC streamer inception criterion can be obtained by transformation of Eq. 4.18:

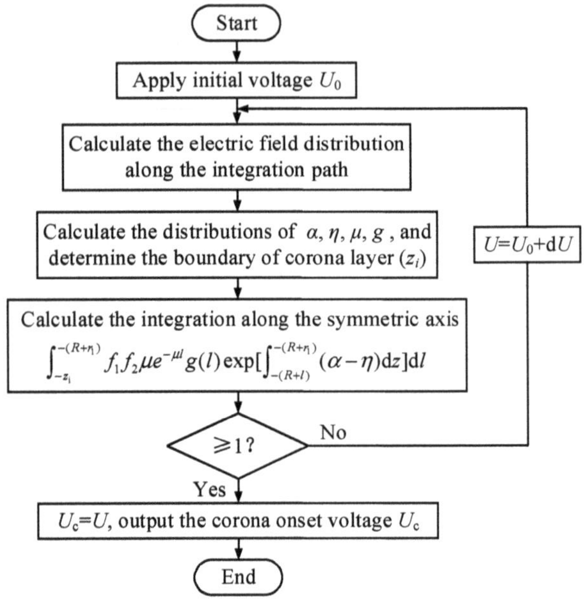

Fig. 4.6 Flow chart of corona onset voltage calculation based on the photoionization model

$$\int_{-z_i}^{-(R+r_1)} f_1 f_2 \mu e^{-\mu l} g(l) \exp\left[\int_{-(R+l)}^{-(R+r_1)} (\alpha - \eta)\mathrm{d}z\right]\mathrm{d}l \geq 1 \tag{4.19}$$

The flow chart of the corona onset voltage calculation by the photoionization model is shown in Fig. 4.6, which is similar with that of the critical charge criterion. Using the photoionization model to calculate the corona onset voltages of the test samples shown in Table 4.1, the results are shown in Table 4.5.

The error indexes of the calculation results by photoionization model are $e_{\mathrm{SSE}} = 109.4686$, $e_{\mathrm{MSE}} = 0.2906$, $e_{\mathrm{MAPE}} = 0.0621$, $e_{\mathrm{MSPE}} = 0.0122$.

4.2.3.3 Empirical and Semi-empirical Formulas

Three empirical and semi-empirical formulas, including the streamer inception electric field criterion [13], Lowke's formula [10] and Ortéga's formula [14, 15] are used to calculate the corona onset voltages of the test samples in Table 4.1.

In engineering, the widely used streamer inception electric field criterion for a rod-plane gap under positive DC voltage is [13]:

$$E_{c+} = 22.8\left(1 + \frac{1}{\sqrt[3]{R}}\right) \tag{4.20}$$

Table 4.5 Positive DC corona onset voltage calculation results of rod-plane electrodes by photoionization model: (**a**) $R = 1.5$ mm, (**b**) $R = 2$ mm, (**c**) $R = 4$ mm, (**d**) $R = 10$ mm

d (cm)	Test value (kV)	Calculated value (kV)	Relative error (%)
a			
1.0	11.28	12.76	13.12
1.5	12.04	13.83	14.87
2.0	12.81	14.45	12.80
3.0	14.33	15.22	6.21
4.0	15.40	15.70	1.95
b			
1.5	15.24	16.12	5.77
2.5	17.02	17.56	3.17
3.5	17.96	18.25	1.61
4.5	18.61	18.88	1.45
5.5	19.93	19.18	−3.76
6.0	20.11	19.36	−3.73
6.5	20.48	19.48	−4.88
7.0	20.66	19.55	−5.37
7.5	20.85	19.67	−5.66
8.0	20.94	19.70	−5.92
8.5	21.20	19.74	−6.89
9.0	21.39	19.76	−7.62
9.5	21.76	19.99	−8.13
c			
1.0	18.14	20.38	12.35
1.5	20.28	23.06	13.71
2.0	22.57	24.70	9.44
2.5	23.63	25.83	9.31
3.0	24.71	26.73	8.17
3.5	26.00	27.44	5.54
4.5	28.00	28.39	1.39
5.5	29.40	29.12	−0.95
6.0	29.89	29.45	−1.47
d			
1.0	25.67	27.91	8.73
2.0	35.53	38.69	8.89
3.0	41.00	44.17	7.73
4.0	44.44	47.61	7.13
6.0	48.44	51.70	6.73
10.0	54.06	55.54	2.74
15.0	57.82	57.85	0.05
20.0	60.39	59.09	−2.15
25.0	62.42	59.82	−4.17

Equation 4.20 is applicable to the standard air density, where R is the rod radius, 0.5 cm $\leq R \leq$ 25 cm.

Lowke et al. [10] deduced the formula to calculate the corona inception field strength of point-plane gaps, which can be expressed as

$$E_c = 25\delta\left(1 + \frac{0.35}{\sqrt{\delta R}} + \frac{0.03}{\delta R}\right) \tag{4.21}$$

where R is the point radius, and δ is the relative air density.

The calculated results of Eqs. 4.20 and 4.21 are both the corona onset electric field strength. By calculating the maximum electric field strength E_m (kV/cm) of this electrode under the applied voltage of 1 kV, the corona onset voltage can be derived as $U_c = E_c/E_m$ (kV).

Ortéga [14, 15] deduced the formula from Peek' law to calculate the corona onset voltage for point-plane geometry, which can be expressed as

$$U_c = \frac{1}{2} R E_0 \log\left(\frac{4d}{R}\right)\left(1 + \frac{0.0436}{\sqrt{R}}\right) \tag{4.22}$$

where R is the point radius, d is the gap length and E_0 is the critical field at which ionization phenomena become significant.

The calculation results of the positive DC corona onset voltage of rod-plane electrodes by the above 3 formulas are shown in Table 4.6.

It can be seen from Table 4.6 that the error indexes of the prediction results by the streamer inception electric field criterion, Lowke's formula and Ortéga's formula are respectively $e_{SSE} = 185.5644$, $e_{MSE} = 0.3784$, $e_{MAPE} = 0.0752$, $e_{MSPE} = 0.0155$; $e_{SSE} = 2462.5211$, $e_{MSE} = 1.3784$, $e_{MAPE} = 0.2728$, $e_{MSPE} = 0.0463$; $e_{SSE} = 664.1624$, $e_{MSE} = 0.7159$, $e_{MAPE} = 0.1502$, $e_{MSPE} = 0.0269$.

4.2.3.4 Comparative Analysis

The errors of prediction results for the positive DC corona onset voltages of rod-plane electrodes obtained by different methods are given in Table 4.7. Here, the results of the photoionization model, the streamer inception electric field criterion, Lowke's formula and Ortéga's formula are calculated under the standard atmospheric condition, the relative air density δ is set as 1. It can be seen that the prediction accuracy of the 6 methods from high to low in order is: SVM model > photoionization model > critical charge criterion > streamer inception electric field criterion > Ortéga's formula > Lowke's formula, which indicates that the prediction method based on electric field features and SVM has the highest accuracy. The e_{MAPE}s of the photoionization model, the critical charge criterion and the streamer inception electric field criterion are in the range of 6–8%, but the absolute percentage errors of some samples exceed 10%, which indicates the applicability of these methods or the parameter values

Table 4.6 Positive DC corona onset voltage calculation results of rod-plane electrodes by empirical and semi-empirical formulas: (**a**) $R = 1.5$ mm, (**b**) $R = 2$ mm, (**c**) $R = 4$ mm, (**d**) $R = 10$ mm

d (cm)	Test value (kV)	Streamer inception electric field criterion		Lowke's formula		Ortéga's formula	
		Calculated value (kV)	Relative error (%)	Calculated value (kV)	Relative error (%)	Calculated value (kV)	Relative error (%)
a							
1.0	11.28	11.26	−0.18	9.01	−20.12	8.49	−24.73
1.5	12.04	12.17	1.08	9.74	−19.10	9.54	−20.76
2.0	12.81	12.72	−0.70	10.18	−20.53	10.29	−19.67
3.0	14.33	13.37	−6.70	10.70	−25.33	11.34	−20.87
4.0	15.40	13.82	−10.26	11.06	−28.18	12.08	−21.56
b							
1.5	15.24	14.44	−5.25	11.29	−25.92	11.57	−24.08
2.5	17.02	15.68	−7.87	12.26	−27.97	13.31	−21.80
3.5	17.96	16.37	−8.85	12.80	−28.73	14.45	−19.54
4.5	18.61	16.85	−9.46	13.17	−29.23	15.31	−17.73
5.5	19.93	17.13	−14.05	13.40	−32.77	15.99	−19.77
6.0	20.11	17.30	−13.97	13.53	−32.72	16.29	−19.00
6.5	20.48	17.37	−15.19	13.58	−33.69	16.56	−19.14
7.0	20.66	17.52	−15.20	13.70	−33.69	16.81	−18.64
7.5	20.85	17.58	−15.68	13.75	−34.05	17.05	−18.23
8.0	20.94	17.63	−15.81	13.79	−34.15	17.27	−17.53
8.5	21.20	17.74	−16.32	13.87	−34.58	17.47	−17.59
9.0	21.39	17.77	−16.92	13.90	−35.02	17.67	−17.39
9.5	21.76	17.86	−17.92	13.96	−35.85	17.85	−17.97
c							
1.0	18.14	18.74	3.31	14.20	−21.73	15.26	−15.88
1.5	20.28	21.11	4.09	15.99	−21.15	17.95	−11.49
2.0	22.57	22.62	0.22	17.13	−24.10	19.85	−12.05
2.5	23.63	23.68	0.21	17.94	−24.08	21.33	−9.73
3.0	24.71	24.48	−0.93	18.54	−24.97	22.54	−8.78
3.5	26.00	25.11	−3.42	19.02	−26.85	23.56	−9.39
4.5	28.00	26.03	−7.04	19.72	−29.57	25.23	−9.89
5.5	29.40	26.69	−9.22	20.21	−31.26	26.56	−9.66
6.0	29.89	26.95	−9.84	20.41	−31.72	27.13	−9.23

(continued)

Table 4.6 (continued)

d (cm)	Test value (kV)	Streamer inception electric field criterion		Lowke's formula		Ortéga's formula	
		Calculated value (kV)	Relative error (%)	Calculated value (kV)	Relative error (%)	Calculated value (kV)	Relative error (%)
d							
1.0	25.67	27.42	6.82	20.75	−19.17	22.42	−12.66
2.0	35.53	37.50	5.54	28.37	−20.15	33.64	−5.32
3.0	41.00	42.78	4.34	32.36	−21.07	40.20	−1.95
4.0	44.44	46.09	3.71	34.87	−21.54	44.85	0.92
6.0	48.44	49.85	2.91	37.72	−22.13	51.41	6.13
10.0	54.06	53.69	−0.68	40.62	−24.86	59.67	10.38
15.0	57.82	55.76	−3.56	42.18	−27.05	66.23	14.55
20.0	60.39	56.86	−5.85	43.02	−28.76	70.88	17.37
25.0	62.42	57.58	−7.75	43.57	−30.20	74.49	19.34

Table 4.7 Error indexes of prediction results for positive DC corona onset voltage of rod-plane electrodes obtained by different methods

Errors	SVM model	Critical charge criterion	Photoionization model	Streamer inception electric field criterion	Lowke's formula	Ortéga's formula
e_{SSE}	12.9796	203.7565	109.4686	185.5644	2462.5211	664.1624
e_{MSE}	0.1001	0.3965	0.2906	0.3784	1.3784	0.7159
e_{MAPE}	0.0172	0.0679	0.0621	0.0752	0.2728	0.1502
e_{MSPE}	0.0037	0.0143	0.0122	0.0155	0.0463	0.0269

are limited. Errors of Lowke's and Ortéga's formulas are so large that they are not suitable to the conditions with high computational accuracy.

The comparison between experimental and predicted values of the positive DC corona onset voltages of rod-plane electrodes is shown in Fig. 4.7 [16]. It can be seen that the predicted results of the SVM model are the closest to the experimental results both in values and variation trend. The variation tendency of the corona onset voltage curves obtained by the photoionization model, critical charge criterion, streamer inception electric field criterion, and Lowke's formula are almost the same. They all tend to saturation faster than the experimental values. In most cases, the calculated results of the photoionization model are larger than experimental values. The results of the critical charge criterion and the streamer inception electric field criterion are close to the experimental values under small gap lengths, but lower than the experimental values under larger gap lengths. The predicted results of Lowke's formula are lower than experimental values. In addition, the tendency of the curves

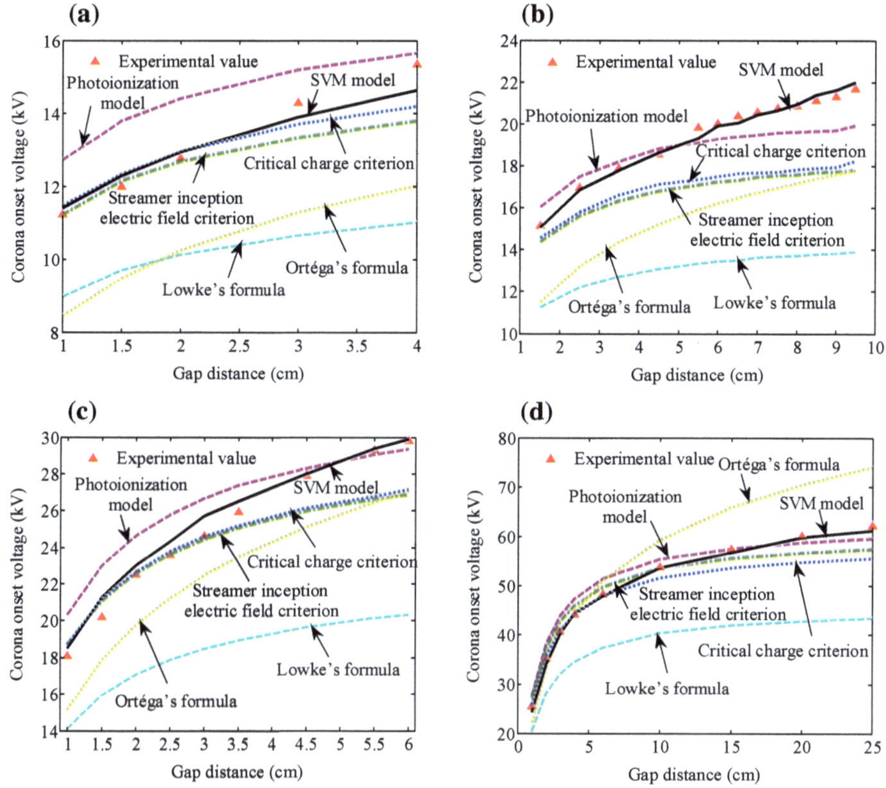

Fig. 4.7 Comparison between predicted and experimental values of the positive DC corona onset voltage: **a** $R = 1.5$ mm, **b** $R = 2$ mm, **c** $R = 4$ mm, **d** $R = 10$ mm

obtained by Ortéga's formula is close to the experimental results when the radius of the hemispherical rod head ranges from 1 to 4 mm, but with much lower values than the test. However, when $R = 10$ mm, the rise rate of the curve is much higher than experimental results, and the values are consistent with the test only when the gap length is within the range from 3 to 5 cm.

The comparison between predicted results of positive DC corona inception electric field strength of rod-plane electrodes obtained by different prediction methods and the experimental values is shown in Fig. 4.8. The gap distance $d = 1$ cm, and the radius of the rod head is respectively 1.5 mm, 2.0 mm, 4.0 mm, and 10.0 mm. It can be seen from Fig. 4.8 that the corona inception field strength obtained by SVM prediction model and the photoionization model, streamer inception critical charge criterion, and streamer inception electric field criterion is very close to the experimental values . SVM model has the highest prediction accuracy. On the other hand, the calculated

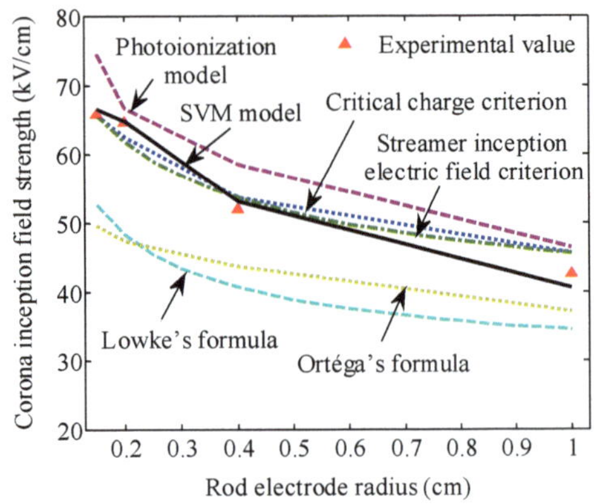

Fig. 4.8 Comparison of predicted and experimental values of positive DC corona inception field strength in rod-plane gaps ($d = 1$ cm)

values of Lowke's formula and Ortéga's formula are much smaller than the test results. Thus, it can be seen that the prediction accuracy of the SVM model is higher than other methods, which verifies the performance of the proposed air insulation prediction method.

4.3 Corona Onset Voltage Prediction of Stranded Conductors

The electromagnetic environment is one of the major technical problems in the design, construction and operation of UHV transmission lines, which is closely related to the corona characteristics of transmission lines. Corona effect on transmission lines may lead to a series of consequences, including corona loss, ion flow, radio interference and audible noise. Corona effect has become one of the decisive factors in the conductor selection of UHV transmission lines, and the corona onset voltage (or inception field strength) is an important basis for corona control.

The aluminium conductor steel reinforced (ACSR) constituted with multistrand aluminium wires and steel wires is the conductor used as high voltage transmission line. The inner steel wires are used to bear the conductor tension, and the outer aluminum wires play the role of current transmission. The initial Peek's formula is only applicable to calculate the corona inception field strength of a smooth cylindrical conductor under ideal conditions. However, there is a great difference of the surface electric field strength between the ACSR and the smooth cylindrical conductor, and therefore their corona onset voltages are quite different. Hence, the surface roughness coefficient m is introduced into the Peek's formula to correct the effect of electric field distortion on the stranded conductor surface. Researchers usually determine the value of m according to experience or specific experimental conditions, which is empirical and of limitations. Therefore, it is necessary to establish a simulation model to directly calculate the surface electric field strength of the stranded conductor, and then predict the corona onset voltage accurately.

4.3.1 Electric Field Analysis of the Stranded Conductor

The calculation model of the stranded conductor is shown in Fig. 4.9. The radiuses of the stranded conductor and a single-stranded aluminum wire are respectively R and r, and the height over ground is H.

The electric field distributions of a stranded conductor with the equivalent radius of 13 mm and a smooth conductor with the radius of 13 mm are calculated. The stranded conductor is composed of a steel core surrounded by 6 stranded aluminium wires. The height of the stranded conductor to the ground is $H = 10$ m, and the applied voltage is 100 kV. The surface and spatial electric field distributions of the smooth conductor and the stranded conductor are shown in Fig. 4.10a, b, respectively.

It can be seen from Fig. 4.10 that the maximum electric field strength on the stranded conductor surface is 15.195 and 10.952 kV/cm on the smooth conductor surface. Their ratio is 1.387, which is approximately consistent with 1.4 obtained in [17].

The electric field strength at different distances from the conductor surface along z axis is shown in Fig. 4.11. The electric field strength near the stranded conductor surface is obviously larger than that of the smooth conductor, and then they will

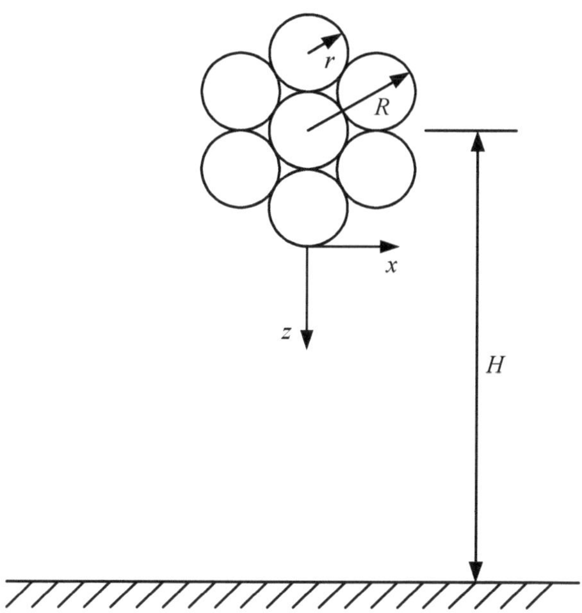

Fig. 4.9 Electric field calculation model of stranded conductor

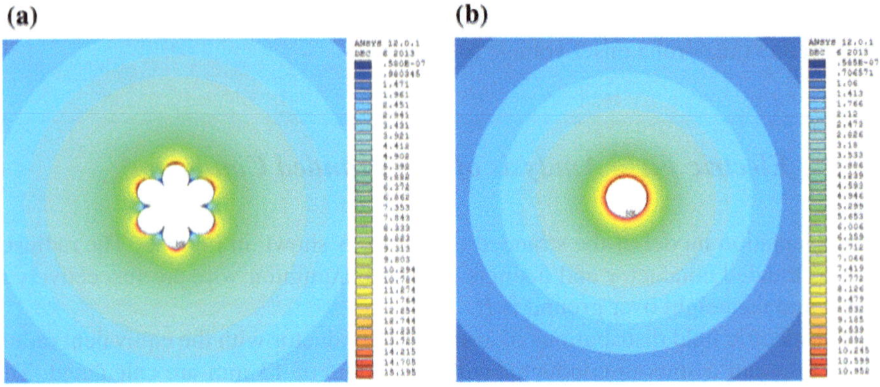

Fig. 4.10 Spatial electric field distributions of the smooth conductor and stranded conductors: **a** smooth conductor, **b** stranded conductor

Fig. 4.11 Electric field strength at different distances from the conductor surface

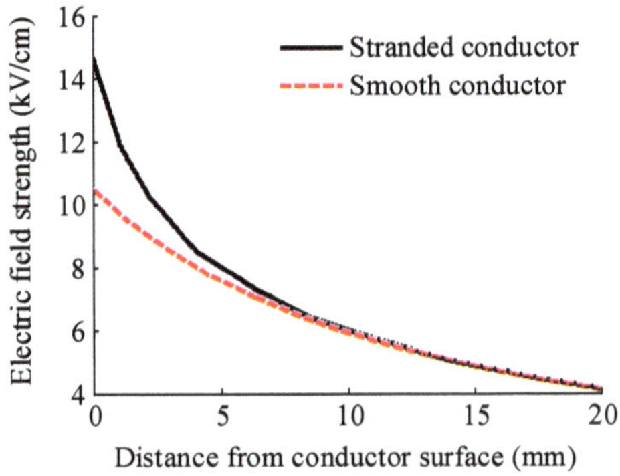

tend to be the same. The ionization zone is the area with high field strength, for the conductor, it is generally within several millimeters from the conductor surface. In this zone, the electric field strength of the stranded conductor is higher than that of the smooth conductor, therefore the corona onset voltage of the stranded conductor is lower than that of the smooth conductor.

The electric field distribution around the surface of the stranded conductor is shown in Fig. 4.12. It can be seen that there are 6 peak values of the field strength, which approximately appear at the angle of $60° \times k$ ($k = 0, 1, 2, 3, 4, 5$), and the electric field strength at the connection point between two strands of aluminium wire is almost 0.

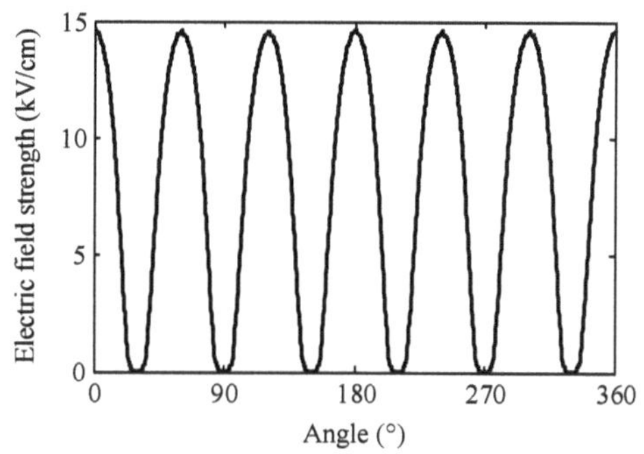

Fig. 4.12 Electric field distribution around the surface of a stranded conductor

4.3.2 *Corona Onset Voltage Prediction of Single Stranded Conductors*

4.3.2.1 Training and Test Sample Set

The negative DC corona onset voltages of stranded conductors measured at different heights over ground H are shown in Table 4.8 [18]. The equivalent radiuses of the stranded conductors are $R = 1$, 1.25 and 1.5 mm. In Table 4.8, the italic bold data are selected as training samples, and the other data are test samples.

4.3.2.2 SVM Prediction Results and Analysis

Under the sense of 3-CV, the improved GS algorithm is used to optimize the SVM parameters, and the correlation analysis method is used to reduce the electric field feature dimensions, so as to analyze the influence of different feature dimensions on the prediction results. For training samples, their corona onset voltages are also predicted by the SVM model.

Table 4.8 Measured values of negative DC corona onset voltage of stranded conductor

H (m)	R (mm)		
	1	1.25	1.5
0.11	27.04	***30.47***	33.66
0.16	28.91	33.29	36.86
0.21	30.56	***35.37***	38.83
0.26	31.51	36.43	40.67
0.36	33.66	***38.51***	43.64
0.46	34.70	40.22	45.03
0.56	***35.50***	***41.56***	***46.84***

Table 4.9 Negative DC corona onset voltage prediction results of stranded conductor under different feature dimensions: (**a**) $R = 1$ mm, (**b**) $R = 1.25$ mm, (**c**) $R = 1.5$ mm

H (m)	Test value (kV)	29-d		13-d		5-d	
		Predicted value (kV)	Relative error (%)	Predicted value (kV)	Relative error (%)	Predicted value (kV)	Relative error (%)
a							
0.11	27.04	26.78	−1	26.50	−2	26.23	−3
0.16	28.91	28.91	0	28.62	−1	28.33	−2
0.21	30.56	30.25	−1	30.25	−1	29.64	−3
0.26	31.51	31.51	0	31.19	−1	30.88	−2
0.36	33.66	32.99	−2	32.99	−2	32.65	−3
0.46	34.70	34.35	−1	34.35	−1	34.00	−2
0.56	35.50	35.50	0	35.15	−1	35.15	−1
b							
0.11	30.47	30.77	1	30.77	1	30.47	0
0.16	33.29	33.62	1	33.29	0	33.29	0
0.21	35.37	35.37	0	35.72	0	35.02	−1
0.26	36.43	36.79	1	36.43	0	36.43	0
0.36	38.51	38.90	1	38.51	0	38.90	1
0.46	40.22	40.22	0	40.22	0	40.22	0
0.56	41.56	41.56	0	41.56	0	41.56	0
c							
0.11	33.66	34.33	2	34.67	3	34.33	2
0.16	36.86	37.23	1	37.60	2	37.60	2
0.21	38.83	39.61	2	39.61	2	39.61	2
0.26	40.67	41.08	1	41.08	1	41.48	2
0.36	43.64	43.64	0	43.64	0	44.08	1
0.46	45.03	45.48	1	45.48	1	45.93	2
0.56	46.84	46.84	0	46.84	0	47.31	1

After feature dimension reduction by correlation analysis method, the negative DC corona onset voltage prediction results of the stranded conductors are shown in Table 4.9. 29-d is all of the electric field features, 13-d is reduced from 29-d by rejecting the features whose correlation coefficient with the corona onset voltage is less than 0.3, and 5-d is reduced from 13-d by rejecting the features whose cross correlation coefficient is greater than 0.9.

The optimal parameters and error indexes of the prediction model under different feature dimensions are shown in Table 4.10. It can be seen that the prediction accuracy of different feature dimensions is 29-d > 13-d > 5-d. Generally, the prediction results under each feature dimension are of high accuracy.

Table 4.10 Optimal parameters and errors of SVM prediction model under different feature dimensions

Parameters and errors	29-d	13-d	5-d
C	8	8	8
γ	0.0045	0.0078	0.0044
e_{SSE}	2.7851	3.8956	7.6131
e_{MSE}	0.0795	0.0940	0.1314
e_{MAPE}	0.0076	0.0090	0.0143
e_{MSPE}	0.0022	0.0027	0.0038

4.3.3 Comparison with Other Prediction Methods

4.3.3.1 Photoionization Model

If the conductor is applied negative DC voltage, when the electric field strength near the conductor surface exceeds a certain value, the free electrons will collide with air molecules and produce initial electron avalanche, as shown in Fig. 4.13. When the photons reach the cathode surface produces at least one photoelectron, a new secondary electron avalanche can be formed, and thus the negative DC corona will incept.

As shown in Fig. 4.13, if there is a free electron at the cathode surface, during its movement towards the ground, it will collide with air molecules and produce the initial electron avalanche. When the initial electron avalanche develops to the position of z, the contained electron number $N_e(z)$ is

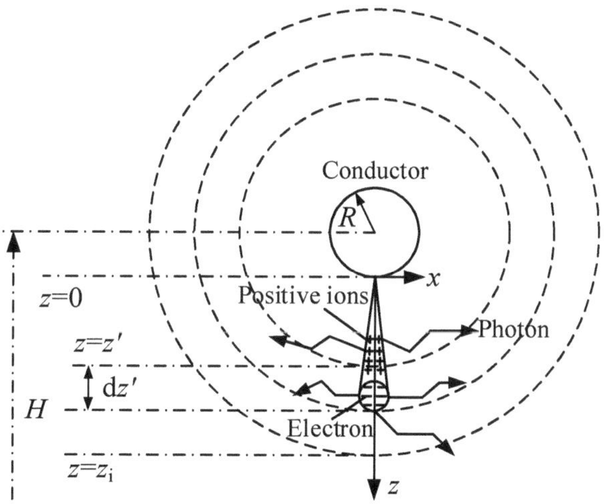

Fig. 4.13 Schematic diagram of negative corona inception on a conductor

$$N_e(z) = \exp\left(\int_0^z (\alpha(z') - \eta(z'))dz'\right) \tag{4.23}$$

Within the distance of Δz, the number of the photons radiated along with the collision ionization of these free electrons is

$$\Delta n_{ph}(z) = \alpha^*(z) N_e(z) \Delta z \tag{4.24}$$

where $\alpha^*(z)$ is the photon generation rate, which is proportional to the ionization coefficient $\alpha(z)$ [19, 20], and the proportionality coefficient is k, that is

$$\alpha^*(z) = k\alpha(z) \tag{4.25}$$

A portion of the produced photons will be absorbed by air. This effect can be characterized by the photo absorption coefficient μ and geometry factor $g(z)$. $g(z)$ is used to consider the disappeared photons in the electrode which are not absorbed by air. The number of photons that can reach the cathode surface is

$$\Delta n^s_{ph} = k\alpha(z) N_e(z) \Delta z g(z) e^{-\mu z} \tag{4.26}$$

In the ionization region from $z = -z_i$ to $z = 0$, when the photons reaching the cathode surface produce at least one photoelectron, a secondary electron avalanche will be formed, and therefore the negative DC corona can be self-sustained. The self-sustained condition can be expressed as

$$N_{eph} = \gamma_{ph} \int_0^{z_i} \alpha(z) g(z) \exp(-\mu z + \int_0^z (\alpha(z') - \eta(z'))dz')dz \geq 1 \tag{4.27}$$

The voltage under which this equality holds is the negative DC corona onset voltage. N_{eph} is the number of photoelectrons on the cathode surface, and γ_{ph} is the photoelectron emission coefficient of the cathode surface which contains the proportionality coefficient k.

The parameters of the photoionization model for corona inception on the negative conductor are calculated as follows.

1. Ionization coefficient α and attachment coefficient η

In order to make the corona inception criterion applicable to different atmospheric conditions, the relationships between each parameter in the criterion and the relative air density δ must be established. The formulas to calculate the ionization coefficient α and the attachment coefficient η are respectively [21, 22]

$$\frac{\alpha}{\delta} = \begin{cases} 3632\exp\left(-168.0\frac{\delta}{E}\right), & 19.0 \leq \frac{E}{\delta} \leq 45.6 \\ 7358\exp\left(-200.8\frac{\delta}{E}\right), & 45.6 \leq \frac{E}{\delta} \leq 182.4 \end{cases} \tag{4.28}$$

$$\frac{\eta}{\delta} = 9.865 - 0.541\frac{E}{\delta} + 1.145 \times 10^{-2}\left(\frac{E}{\delta}\right)^2 \quad (4.29)$$

where α and η are in cm^{-1}, E is in kV/cm.

2. Photon absorption coefficient μ

The photon absorption coefficient of air μ is assumed to be proportional to the relative air density, that is,

$$\mu = \delta\mu_0 \quad (4.30)$$

where $\mu_0 = 6\ \mathrm{cm}^{-1}$ [21].

3. Geometry factor $g(z)$

According to Aleksandrov's method [23], the geometry factor $g(z)$ can be decomposed into the product of the radial component $g_{\mathrm{rad}}(z)$ and the axial component $g_{\mathrm{ax}}(z)$, that is,

$$g(z) = g_{\mathrm{rad}}(z)g_{\mathrm{ax}}(z) \quad (4.31)$$

For negative conductor, $g_{\mathrm{rad}}(z) \neq g_{\mathrm{ax}}(z)$, the radial component $g_{\mathrm{rad}}(z)$ and the axial component $g_{\mathrm{ax}}(z)$ are calculated by [24]:

$$g_{\mathrm{rad}}(z) = \frac{1}{\pi e^{-\mu(z-R)}} \int_0^{\sin^{-1}(R/z)} e^{-\mu\left(z\cos\psi_1 - \sqrt{R^2 - z^2\sin^2\psi_1}\right)}\,\mathrm{d}\psi_1 \quad (4.32)$$

$$g_{\mathrm{ax}}(z) = \frac{1}{\pi e^{-\mu(z-R)}} \int_0^{\frac{\pi}{2}} e^{-\mu(z-R)/\cos\psi_2}\,\mathrm{d}\psi_2 \quad (4.33)$$

4. Surface photoelectron emission coefficient γ_{ph}

For negative corona, γ_{ph} is usually set as 3×10^{-3} [21].

Using the above photoionization model to calculate the corona onset voltages of the strand conductors under different R and H, as shown in Table 4.8, the results are shown in Table 4.11, where σ is the relative error between the calculated value and the test value of the corona onset voltage. The error indexes are $e_{\mathrm{SSE}} = 85.4880$, $e_{\mathrm{MSE}} = 0.4403$, $e_{\mathrm{MAPE}} = 0.0561$, $e_{\mathrm{MSPE}} = 0.0125$.

4.3.3.2 Critical Charge Criterion

The critical charge criterion for streamer inception, described in Sect. 4.2.3.1 is used to calculate the corona onset voltages of the stranded conductors. The research

Table 4.11 Negative DC corona onset voltage calculation results of stranded conductors by photoionization model

H (m)	$R = 1$ mm			$R = 1.25$ mm			$R = 1.5$ mm		
	Test value (kV)	Calculated value (kV)	σ (%)	Test value (kV)	Calculated value (kV)	σ (%)	Test value (kV)	Calculated value (kV)	σ (%)
0.11	27.04	28.78	6.43	30.47	32.78	7.58	33.66	36.38	8.08
0.16	28.91	30.91	6.92	33.29	35.19	5.71	36.86	39.28	6.57
0.21	30.56	32.33	5.79	35.37	36.95	4.47	38.83	41.27	6.28
0.26	31.51	33.54	6.44	36.43	38.37	5.33	40.67	42.89	5.46
0.36	33.66	35.27	4.78	38.51	40.45	5.04	43.64	45.33	3.87
0.46	34.70	36.64	5.59	40.22	42.04	4.53	45.03	47.14	4.69
0.56	35.50	37.72	6.25	41.56	43.31	4.21	46.84	48.61	3.78

results in [17, 22] demonstrate that the critical charge N_{crit} should be set as 3500 for coaxial smooth conductor with the radius ranging from 0.05 to 2 cm. The radius of the stranded conductor in Table 4.8 ranges from 1 to 1.5 mm, which are within the application scope of $N_{crit} = 3500$.

Using the critical charge criterion with $N_{crit} = 3500$ to calculate the corona onset voltages of the stranded conductors with different values of R and H, the calculation results are shown in Table 4.12. The error indexes are $e_{SSE} = 66.0407$, $e_{MSE} = 0.3870$, $e_{MAPE} = 0.0495$, $e_{MSPE} = 0.0111$.

4.3.3.3 Peek's Formula

On the basis of Peek's formula, an empirical formula used to calculate the negative DC corona onset electric field strength was given in [25, 26], that is

$$E_c = 31\left(1 + \frac{0.308}{\sqrt{\delta R}}\right) \tag{4.34}$$

where δ is the relative air density, R is the conductor radius.

Peek's formula can calculate the corona onset electric field strength of the conductor without complex numerical calculations, and the corona onset voltage can be obtained by electric field calculations. The corona onset voltage of the stranded conductor can be calculated by the modified Peek's formula with the surface roughness coefficient m. The definition of m is

$$m = \frac{U_{cst}}{U_{csm}} \tag{4.35}$$

where U_{cst} and U_{csm} are respectively the corona onset voltages of the stranded conductor and the smooth conductor.

For new conductor, the value of m is usually set as 0.8–0.9. Here, the modified Peek's formula with $m = 0.85$ is used to calculate the corona onset voltages of the stranded conductors, and the calculation results are shown in Table 4.13. The error indexes are $e_{SSE} = 3.8345$, $e_{MSE} = 0.0932$, $e_{MAPE} = 0.0106$, $e_{MSPE} = 0.0028$.

4.3.3.4 Comparative Analysis

Table 4.14 shows the error indexes of the corona onset voltage prediction results for stranded conductors by different methods. It can be seen that the predicted results of the SVM model are of the highest accuracy, and the e_{MAPE} is only 0.81%. The modified Peek's formula with $m = 0.85$ also has very good calculation accuracy, but the error indexes are a bit higher than those of the SVM model. The results indicate that the roughness coefficient m of the stranded conductor is close to 0.85, but this does not mean that it is also suitable for other stranded conductors. The calculation

Table 4.12 Negative DC corona onset voltage calculation results of stranded conductors by critical charge criterion for streamer inception

H (m)	$R = 1$ mm			$R = 1.25$ mm			$R = 1.5$ mm		
	Test value (kV)	Calculated value (kV)	σ (%)	Test value (kV)	Calculated value (kV)	σ (%)	Test value (kV)	Calculated value (kV)	σ (%)
0.11	27.04	28.65	5.95	30.47	32.53	6.76	33.66	36.11	7.28
0.16	28.91	30.77	6.43	33.29	35.03	5.23	36.86	38.89	5.51
0.21	30.56	32.19	5.33	35.37	36.78	3.99	38.83	40.97	5.51
0.26	31.51	33.40	6.00	36.43	38.09	4.56	40.67	42.48	4.45
0.36	33.66	35.11	4.31	38.51	40.26	4.54	43.64	44.92	2.93
0.46	34.70	36.48	5.13	40.22	41.84	4.03	45.03	46.71	3.73
0.56	35.50	37.55	5.77	41.56	43.11	3.73	46.84	48.16	2.82

Table 4.13 Negative DC corona onset voltage calculation results of stranded conductors by Peek's law

H (m)	$R = 1$ mm			$R = 1.25$ mm			$R = 1.5$ mm		
	Test value (kV)	Calculated value (kV)	σ (%)	Test value (kV)	Calculated value (kV)	σ (%)	Test value (kV)	Calculated value (kV)	σ (%)
0.11	27.04	27.50	1.70	30.47	31.17	2.30	33.66	34.56	2.67
0.16	28.91	29.49	2.01	33.29	33.54	0.75	36.86	37.29	1.17
0.21	30.56	30.93	1.21	35.37	35.25	−0.34	38.83	39.26	1.11
0.26	31.51	32.06	1.75	36.43	36.59	0.44	40.67	40.81	0.34
0.36	33.66	33.77	0.33	38.51	38.62	0.29	43.64	43.15	−1.12
0.46	34.70	35.06	1.04	40.22	40.15	−0.17	45.03	44.92	−0.24
0.56	35.50	36.09	1.66	41.56	41.37	−0.46	46.84	46.33	−1.09

Table 4.14 Error indexes of the corona onset voltage prediction results for stranded conductors by different methods

Errors	SVM model	Photoionization model	Critical charge criterion	Peek's formula
e_{SSE}	2.7851	85.4880	66.0407	3.8345
e_{MSE}	0.0795	0.4403	0.3870	0.0932
e_{MAPE}	0.0076	0.0561	0.0495	0.0106
e_{MSPE}	0.0022	0.0125	0.0111	0.0028

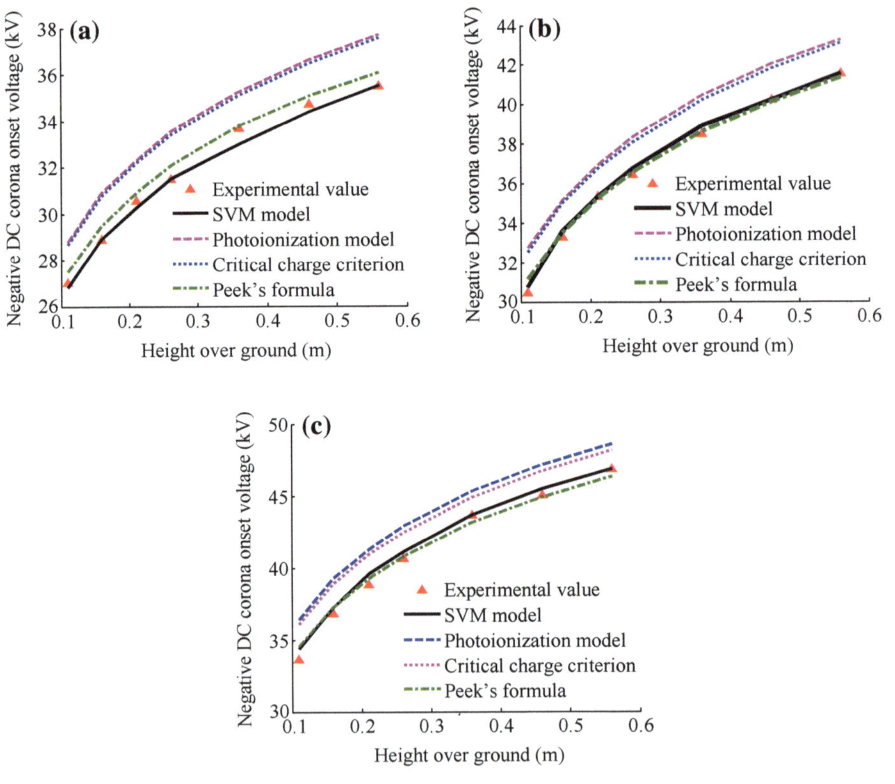

Fig. 4.14 Comparison between the calculated and experimental corona onset voltages of negative DC stranded conductors: **a** $R = 1$ mm, **b** $R = 1.25$ mm, **c** $R = 1.5$ mm

results based on the photoionization model have the largest errors, and the e_{MAPE} is up to 5.61%. The calculation results of the critical charge criterion also have large errors, but a bit better than the photoionization model.

The comparison between the calculated and experimental corona onset voltages of the negative DC stranded conductors is shown in Fig. 4.14, which also verifies the accuracy of the proposed SVM model.

4.4 DC Corona Onset Voltage Prediction of Valve Hall Fittings

The converter station is the hub of the EHV and UHV DC transmission projects. High voltage fittings are important components of the equipment used in converter valve hall, which are used for connection, voltage equalizing and shielding. These fittings are of vital importance to ensure the safe operation of the equipment in the valve hall, and they are essential components to guarantee the reliable operation of the whole DC transmission system. It is of great significance to properly select the valve hall fittings.

One of the main goals for the design of valve hall fittings is to control the surface electric field strength of the fitting within the corona inception field strength, thus to prevent corona inception, so as to avoid the electromagnetic environment problems and the aging of insulation equipment caused by corona effects. The corona inception field strength is an important basis for corona control, which was mostly determined by Peek's law in previous studies [27]. However, Peek's law was derived from a cylindrical structure, and whether it can be used for circular or spherical electrodes remains to be studied. In this section, the DC corona onset voltages of the shielding sphere electrodes used in valve hall were measured, and the corona onset voltages are predicted respectively by the proposed SVM model and the photoionization model. Combing with the electric field calculation results, the fitting formulas are given to calculate the corona inception field strength.

4.4.1 Corona Tests

Corona inception tests of valve hall shielding spheres were carried out at the high voltage laboratory in the UHV DC test base of State Grid Corporation of China (SGCC) in Beijing, China. The test power supply is a ±1800 kV/0.2 A DC voltage generator. The schematic diagram and test arrangement of DC corona inception test are shown in Fig. 4.15.

The diameter of the sphere electrode ranges from 50 to 90 cm, and the plane electrode is a grounded iron sheet rolling out from the vertical of the sphere canter, with the size of 8.87 m × 8.35 m. The gap length is 3 m. Because the grounded plate is not an ideal infinite plate, there may be some errors. However, the error can be neglected since the grounded plate is far from the sphere surface.

Corona tests were carried out according to IEC 61284 [28]. The corona inception phenomenon was observed by an ultraviolet imager. Corona is considered to incept when the photon counting per unit time exceeds 100. Four times of tests were conducted for each sphere diameter. During the test, the atmospheric pressure ranged from 99.91 to 101.41 kPa and the temperature ranged from 15.4 to 17.0 °C, while the relative humidity ranged from 23.6 to 49.3%. Test results of the positive and negative DC corona onset voltages of the shielding sphere-plane gaps are shown in

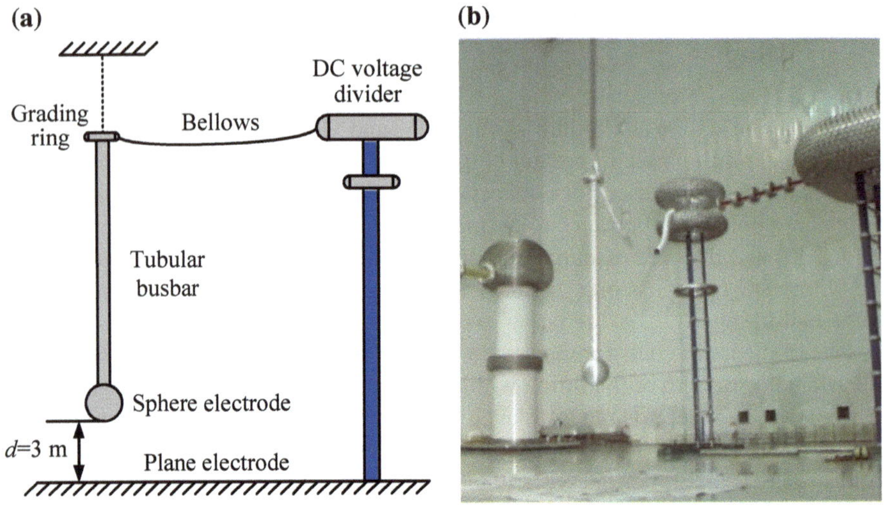

Fig. 4.15 Schematic and arrangement diagram of DC corona test for valve hall shielding sphere: **a** schematic diagram, **b** test arrangement

Table 4.15 Test results of positive and negative DC corona onset voltages of the shielding sphere electrodes

Sphere diameter (cm)	Positive corona onset voltage (kV)			Negative corona onset voltage (kV)		
	U_{cmax}	U_{cmin}	U_{cav}	U_{cmax}	U_{cmin}	U_{cav}
50	698	581	649	598	595	597
60	759	708	726	650	580	607
70	820	760	783	760	710	740
80	1000	970	985	830	810	820
90	1100	1050	1075	870	850	860
100	–	–	–	905	900	902

Table 4.15 without any atmospheric correction, where U_{cmax} and U_{cmin} are respectively the maximum and minimum measured value of the corona onset voltage, and U_{cav} is the average value.

4.4.2 *Corona Onset Voltage Prediction*

The SVM model and the photoionization model are applied to predict the DC corona onset voltages of the shielding sphere electrodes and compare with the experimental results shown in Table 4.15. For positive DC corona onset voltage prediction, the

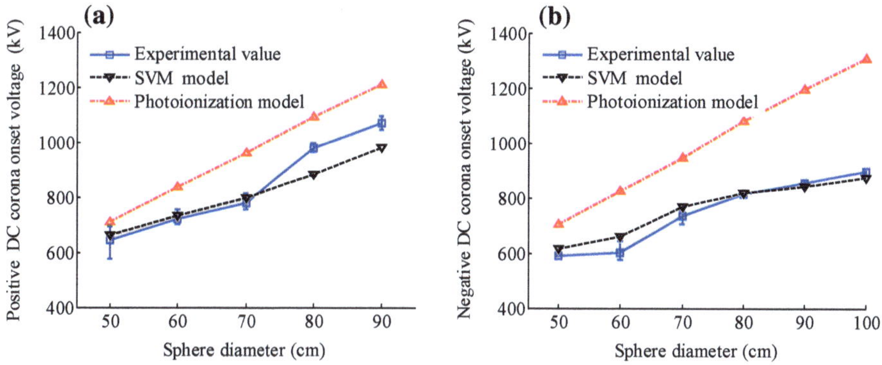

Fig. 4.16 Comparison between predicted and experimental corona onset voltages of valve hall shielding spheres: **a** positive, **b** negative

SVM model was trained by the eight training samples of rod-plane gaps presented in Table 4.1. The input dimension of the prediction model is 29-d. After optimized by the improve GS algorithm, the penalty factor $C = 181.019$ and the kernel parameter $\gamma = 0.0769$. The comparison between the predicted and experimental positive DC corona onset voltage is shown in Fig. 4.16a.

For negative DC corona onset voltage prediction, the SVM model was trained by the training samples of stranded conductors shown in Table 4.8. The input dimension of the prediction model is also 29-d. The optimal SVM parameters are $C = 128$, $\gamma = 0.20306$. The comparison between the predicted and experimental negative DC corona onset voltage is shown in Fig. 4.16b.

A finite element model was established to simulate the valve hall shielding sphere-plane gap and calculate the maximum values of the sphere surface field strength. Combining with the finite element calculation results, the experimental and predicted corona inception field strength of the shielding sphere electrode can be calculated according to the experimental and predicted corona onset voltages. The positive and negative DC corona inception field strengths are shown in Table 4.16.

4.4.3 Result Analysis and Discussions

The following conclusions can be drawn from the experimental corona onset voltages of the valve hall shielding spheres and the comparisons with the predicted results [29].

1. The positive DC corona onset voltage is higher than the negative DC corona inception voltage, whether the experimental value or the predicted value, which is consistent with the gas discharge theory.

Table 4.16 Positive and negative DC corona inception field strengths of the shielding sphere electrodes

Sphere diameter (cm)	Positive corona inception field strength (kV/cm)			Negative corona inception field strength (kV/cm)		
	Test value	SVM model	Photoionization model	Test value	SVM model	Photoionization model
50	27.24	27.78	29.84	25.06	25.81	29.75
60	25.58	25.84	29.71	21.39	23.31	29.24
70	23.86	24.33	29.40	22.55	23.45	28.99
80	26.50	23.77	29.55	22.06	22.06	29.23
90	25.94	23.66	29.31	20.75	20.34	28.99
100	–	–	–	19.77	19.18	28.78

2. The experimental values are with large dispersion, and the dispersion of the positive polarity is larger than that of the negative polarity. For the positive DC corona, the onset voltages of the shielding spheres with the diameters from 80 to 90 cm have an obvious mutation compared with that of from 50 to 70 cm, which results in higher corona inception field strength of the shielding sphere with larger diameter. This phenomenon may be caused by the changes of the environmental conditions during the tests. According to the test record, the relative humidity is higher and close to 50% during the tests of the shielding spheres with the diameters of 50–70 cm, but the relative humidity is less than 25% for the spheres with the diameters of 80–90 cm.
3. Whether positive or negative corona, the SVM prediction results of the corona onset voltage and field strength are much closer to the experimental values than those of the photoionization model, which are much higher than experimental values. In the variation trend of the corona onset voltage and field strength, the SVM prediction results are smoother than the experimental results, which is more in line with the theoretical analysis results. Some parameters in the photoionization model, like the ionization coefficient, the attachment coefficient and the photon absorption coefficient, are not unified, and they are affected by atmospheric parameters such as air pressure and humidity. This may be the reason for large calculation errors of the photoionization model.

The relation equations between the positive and negative DC corona inception field strength and the sphere diameter can be deduced by fitting the experimental data using the least square method:

$$\begin{cases} E_{\text{ct+}} = 71.26D^{-0.2336} & R^2 = 0.5093 \\ E_{\text{ct}-} = 86.57D^{-0.3213} & R^2 = 0.9667 \end{cases} \tag{4.36}$$

where $E_{\text{ct+}}$ and $E_{\text{ct}-}$ are respectively experimental values of the positive and negative corona inception field strength in kV/cm, and D is the sphere diameter in cm.

Table 4.17 Assessment of DC corona inception electric field strength of valve hall shielding spheres

Diameter (cm)	Fitted value (kV/cm)				Safe value (kV/cm)		Actual value (kV/cm)	Corona inception or not
	Positive		Negative					
	Test	SVM	Test	SVM	Test	SVM		
120	23.29	21.28	18.59	18.33	12.87	12.69	7.56	No
160	21.78	19.58	16.95	16.37	11.73	11.33	5.86	No

By the same way, the relation equations between the positive and negative DC corona inception field strength and the sphere diameter can be fitted according to the predicted results, which are expressed as

$$\begin{cases} E_{\mathrm{cp}+} = 85.26D^{-0.2899} & R^2 = 0.9373 \\ E_{\mathrm{cp}-} = 119.3D^{-0.3913} & R^2 = 0.9356 \end{cases} \tag{4.37}$$

where $E_{\mathrm{cp}+}$ and $E_{\mathrm{cp}-}$ are respectively experimental values of the positive and negative corona inception field strength in kV/cm.

Equations 4.36 and 4.37 were applied to calculate the corona inception field strength of the shielding spheres used in a ±660 kV DC convertor station valve hall. The diameters of the spheres are 120 and 160 cm respectively. According to the operation condition of the valve hall shielding spheres, the negative DC corona inception field strength is taken as the control standard. Considering the craft of the shielding spheres, which are composed of two hemispheres, the control value of the corona inception field strength can be obtained by considering the correction factor, set as 0.9, and margin coefficient, set as 1/1.3, in which 1.3 is the overload multiples. Meanwhile, the actual surface electric field strength of the valve hall shielding sphere under operating condition can be calculated by 3D FEM. Whether corona will incept on the shielding sphere can be judged by comparing the safe value and the finite element simulation result, as shown in Table 4.17.

It can be seen from Table 4.17 that the fitted and safe values of the corona inception field strength extrapolated from the predicted results of the SVM model is about 1–4% lower than the values extrapolated from test results, which indicates that the safe value extrapolated from experimental data is more strict. The finite element calculation results indicate that the maximum electric field strength are respectively 7.56 kV/cm and 5.86 kV/cm on the surface of the shielding spheres with the diameters of 120 cm and 160 cm, which are lower than the safe value of the corona inception field strength deduced from experimental or predicted results. Hence, it can be seen that the corona will not incept on the shielding spheres in the operating environment of the valve hall.

4.5 Brief Summary

In this chapter, the air insulation prediction model is used to predict the positive DC corona onset voltages of the rod-plane gaps, the negative DC corona onset voltages of the stranded conductors and the DC corona onset voltages of the valve hall shielding spheres. The conclusions can be drawn as follows:

1. Compared with the critical charge criterion, the photoionization model, the streamer inception electric field criterion, Ortéga's formula and Lowke's formula, the prediction method proposed in this book has higher prediction accuracy for positive DC corona onset voltages of rod-plane electrodes. The e_{MAPE} of the prediction results of the 36 test samples is only 1.72%, which verifies the validity and superiority of the proposed method.
2. For the negative DC corona onset voltage prediction of the single stranded conductor with $R = 1$–1.5 mm and $H = 0.11$–0.56 m, the e_{MAPE} of the predicted results by the method based on electric field features and SVM is only 0.76% compared with the experimental values, which has higher prediction accuracy than the critical charge criterion, the photoionization model and the Peek's formula.
3. Using the proposed method to predict the corona onset voltage and inception field strength of valve hall shielding spheres, the predicted results are in good agreement with the experimental values, and the dispersion of the test results can be overcome, which is more conducive to extrapolate the corona inception field strength of the shielding spheres with other diameters. Whether positive or negative corona, the SVM prediction values of the corona onset voltage and field strength are much closer to the experimental results than those calculated by photoionization model.
4. According to the formulas fitted by experimental and predicted values, and considering the margin coefficient, the control values of the corona inception field strength are calculated for the shielding spheres used in a ±660 kV DC converter station valve hall. Compared with the electric field FEM calculation results, it is proved that the corona will not incept on the shielding sphere in the operating environment of the valve hall. The method proposed in this book for corona onset voltage and inception field strength prediction is possible to provide an effective basis for corona control of valve hall fittings.

References

1. Guan GZ (2003) Fundamentals of high voltage engineering. China Electric Power Press, Beijing
2. Liu ZY (2005) Ultra-high voltage grid. China Economic Publishing House, Beijing
3. Yan Z, Zhu DH (2007) High voltage insulation technology, 2nd edn. China Electric Power Press, Beijing
4. Peek FW (1929) Dielectric phenomena in high voltage engineering. McGraw-Hill, New York

5. Lings R (2005) EPRI AC transmission line reference book-200 kV and above, 3rd edn. Electric Power Research Institute, Palo Alto, USA
6. Nasser E, Heiszler M (1974) Mathematical-physical model of the streamer in non-uniform fields. J Appl Phys 45(8):3396–3401
7. Tsuneysu I, Nisijima K, Izawa Y (1992) Flashover phenomena in positive rod-to-plane air gaps under impulse and DC voltages. Electr Eng Jpn 112(6):20–32
8. Isa H, Sonoi Y, Hayashi N (1991) Breakdown process of a rod-to-plane gap in atmospheric air under dc voltage stress. IEEE Trans Dielectr Electr Insul 26(2):291–299
9. Reather H (1964) Electron avalanches and breakdown in gases. Butterworth, London
10. Lowke JJ, Alessandro FD (2003) Onset corona fields and electrical breakdown criteria. J Phys D Appl Phys 36(21):2673–2682
11. Shu SW, Liu C, Ruan JJ (2015) Comparative of positive DC corona inception criteria for a rod-plane electrode. Eng J Wuhan Univ 48(6):836–847
12. Nasser E, Abou-Seada M (1970) Calculation of streamer thresholds using digital techniques. IEE Conf Pub 70:534–537
13. Abdel-Salam M, Allen NL (1990) Inception of corona and rate of rise of voltage in diverging electric field. IEE Proc A Sci Meas Technol 137(4):217–220
14. Ortéga P (1992) Comportement diélectrique des grands intervalles d'air soumis à des ondes de tension de polarité positive ou negative. Dissertation, Université de Pau
15. Fofana I, Beroual A (1997) A predictive model of positive discharge in long air gaps under pure and oscillating impulse shapes. J Phys D Appl Phys 30(11):1653–1667
16. Qiu ZB, Ruan JJ, Huang DC et al (2016) Prediction study on positive DC corona onset voltage of rod-plane air gaps and its application to the design of valve hall fittings. IET Gener Transm Distrib 10(7):1519–1526
17. Yamazaki K, Olsen RG (2004) Application of a corona onset criterion to calculation of corona onset voltage of stranded conductors. IEEE Trans Dielectr Electr Insul 11(4):674–680
18. Bahy MME, Abouelsaad M, Gawad NA et al (2007) Onset voltage of negative corona on stranded conductors. J Phys D Appl Phys 40(5):3094–3101
19. Meng XB, Bian XM, Chen FL et al (2011) Analysis on negative DC corona inception voltage of stranded conductors. High Volt Eng 37(1):77–84
20. Salam MA (1993) Calculation of corona onset voltage for duct-type precipitators. IEEE Trans Ind Appl 29(2):274–280
21. Sarma MP, Janischewskyj D (1969) DC corona on smooth conductors in air. Steady-state analysis of the ionization layer. Proc IEE 116(1):161–166
22. Phillips DB, Olsen RG, Pedrow PD (2000) Corona onset as a design optimization criterion for high voltage hardware. IEEE Trans Dielectr Electr Insul 7(6):744–751
23. Aleksandrov GN (1956) Physical conditions for the formation of an alternating current corona discharge. Soviet Phys-Tech Phys 1(8):1714–1726
24. Salam MA (1976) Calculating the effect of high temperatures on the onset voltages of negative discharges. J Phys D 9(12):149–154
25. Stockmeyer W (1934) Koronaverluste bei hoher Gleichspannung. Wiss Verofl Siemens 13(2):27–34
26. Ouyang KW (2012) Research on influence factors and calculation method of DC corona onset voltage. Dissertation, North China Electric Power University
27. Ruan JJ, Zhan T, Du ZY et al (2013) Numerical solution of surface electric field in ±800 kV UHVDC converter station valve hall of electric power fittings. High Volt Eng 39(12):2916–2923
28. IEC 61284 (1997) Overhead lines-requirements and tests for fittings
29. Qiu ZB, Ruan JJ, Huang DC et al (2016) Prediction on corona onset voltage of DC conductors and valve hall fittings. Trans China Electro Tech Soc 31(12):80–89

Chapter 5
Power Frequency Breakdown Voltage Prediction of Air Gaps

5.1 Air Gap Breakdown Characteristics Under Steady-State Voltage

The electric field distribution of an air gap has a great influence on its breakdown characteristics. In order to describe the inhomogeneity degree of the electric field, the electric field nonuniform coefficient f is introduced. It is the ratio of the maximum field strength $E_{\max}$ and the average field strength E_{av}.

$$f = E_{\max}/E_{\mathrm{av}} \tag{5.1}$$

$$E_{\mathrm{av}} = U/d \tag{5.2}$$

where U is the applied voltage, and d is the gap distance.

For a uniform electric field, $f = 1$. Generally speaking, the electric field is slightly uneven when $f < 2$, and it is extremely nonuniform when $f > 4$.

The electric field inhomogeneity degree can also be divided qualitatively according to whether the corona discharge can be sustained. If the electric field is nonuniform enough to sustain the corona discharge, then it is called the extremely nonuniform electric field. Although the electric field is not uniform, but it is unable to sustain stable corona discharge, and the gap will breakdown immediately once the discharge is self-sustained, then it is called the slightly uneven electric field. The breakdown process in a slightly uneven electric field is similar to that in a uniform electric field. However, in an extremely nonuniform electric field, stable corona discharge will appear before gap breakdown and the discharge process has significant polarity effects. In addition, leader discharge process will happen for long air gaps.

Air gap discharge process and the characteristics of electric field distribution can be summarized as Fig. 5.1. Air gap discharge process can be divided into the following stages: weak ionization, streamer corona inception, streamer propagation, leader inception, continuous leader propagation, final jump and gap breakdown. The characteristics of the electric field during the discharge process can be divided

Z. Qiu et al., *Air Insulation Prediction Theory and Applications*, Power Systems,
https://doi.org/10.1007/978-981-10-5163-0_5

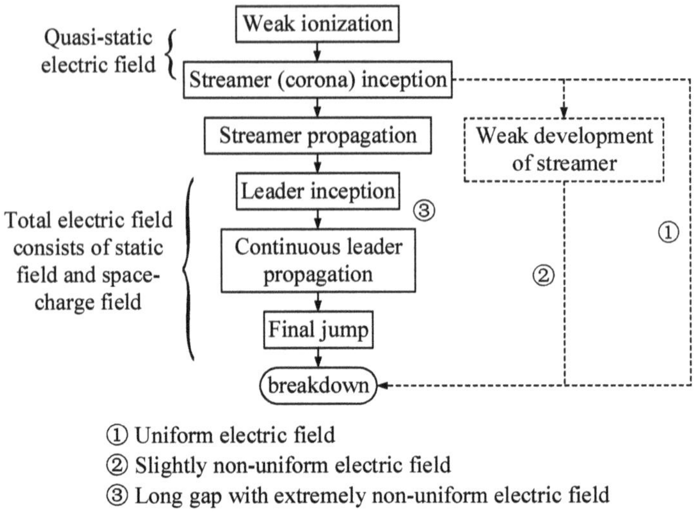

Fig. 5.1 Air gap discharge process and characteristics of the electric field

into two types, one is the quasi-electrostatic field, and the other is the total electric field of the electrostatic field and the space charge field. In a uniform electric field, the gap will breakdown after streamer inception. In a slightly uneven electric field, the gap will breakdown after a weak development of the streamer. Therefore, for air gaps with uniform and slightly uneven electric fields, the electric field during the discharge process can be considered as a quasi-electrostatic field, which is not affected by the spatial electric field generated by corona charges. For air gaps with extremely nonuniform electric field, the electric field is a quasi-electrostatic field before streamer corona inception, and it is a total electric field after streamer corona inception.

5.1.1 Breakdown in Uniform Electric Field

Uniform electric field is rare in engineering. Strictly speaking, only the air gap with infinite parallel plate electrodes has the uniform electric field. The parallel plate electrodes used in engineering are generally with chamfered edges of large curvature radius, so as to eliminate the electrode edge effect.

Corona phenomenon does not exist in a uniform electric field. The gap will breakdown once discharge initiation, so the DC, AC (peak value) and impulse breakdown voltages of an air gap in a uniform electric field are the same, and the dispersion of the breakdown voltage is very small.

According to Paschen's law, the empirical formula to calculate the breakdown voltages (peak value) of air gaps in uniform electric field can be derived as:

$$U_b = 24.22\delta d + 6.08\sqrt{\delta d} \quad (5.3)$$

where the breakdown voltage U_b is in kV, the gap distance d is in cm, and δ is the relative air density.

It can be seen from Eq. 5.3 that the breakdown field strength E_b is about 30 kV/cm in a uniform electric field when $\delta = 1$.

5.1.2 Breakdown in Slightly Uneven Electric Field

The most common electric field in engineering is the nonuniform electric field. According to the characteristics of the discharge phenomena and the discharge process, the nonuniform electric field can be divided into slightly uneven electric field and extremely nonuniform electric field. Under steady-state voltage, the breakdown characteristics in a slightly uneven electric field are similar to that in a uniform electric field. There is no corona before gap breakdown, and the dispersion of the breakdown voltage is not large.

The breakdown voltages of air gaps in slightly uneven electric field are closely related to the electric field inhomogeneity degree. At the same gap distance, the breakdown voltage is higher for more uniform electric field. A typical slightly uneven electric field is the standard sphere gap given in IEC 60052 [1] for high voltage measurement. This is a pair of sphere electrodes with the same diameter, one sphere is applied high voltage, and the other is grounded. The electric field inhomogeneity degree of sphere gaps varies with d/D, where d is the gap distance and D is the sphere diameter. Taking the spheres with the diameter of 5–50 cm as examples, the relation between the steady-state breakdown voltage U_b and the gap distance d is shown in Fig. 5.2.

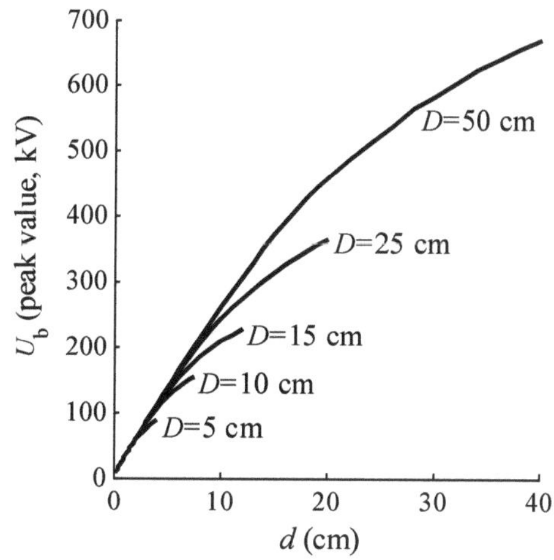

Fig. 5.2 Relationship between sphere gap breakdown voltage and the gap distance (one sphere grounded)

Accurately determination of air gap breakdown voltage in slightly uneven electric field depends on experiments, but there are also some estimation methods. According to Eqs. 5.1 and 5.2, it can be drawn that $U = E_{max}d/f$. As mentioned before, air gap breakdown strength in a uniform electric field is about 30 kV/cm. It can be considered that the gap will breakdown in slightly uneven electric field when E_{max} reaches $E_0 = 30$ kV/cm. Therefore, the breakdown voltage (peak value) can be estimated according to the following formula:

$$U_b = E_0 d/f = 30d/f \tag{5.4}$$

where f can be obtained by static electric field calculation.

In addition, once the streamer formation in a slightly uneven electric field, it will develop through the whole gap and lead to gap breakdown. Hence, the breakdown voltage can also be calculated by empirical formulas and physical models, such as the critical charge criterion, the streamer inception field criterion and the photoionization model.

5.1.3 Breakdown in Extremely Nonuniform Electric Field

Air gaps with extremely nonuniform electric field are the most common in engineering, and the most typical gap types are the rod-plane and rod-rod gaps. Rod-plane gap has the most asymmetric structure and rod-rod gap is completely symmetrical. The breakdown characteristics of other air gaps with extremely nonuniform electric field are between these two kinds of typical gaps.

Under DC voltage, the rod-plane gap breakdown voltage has obvious polarity effect. When the rod electrode is applied positive DC voltage, the breakdown voltage is much lower than that under negative DC voltage, while the rod-rod gap breakdown voltage is between those of different polarities of rod-plane gaps. Test results showed that [2] when the gap distance $d < 300$ cm, the average breakdown field strength of the positive rod-plane air gap is about 4.5 kV/cm, and it is about 10 kV/cm for negative rod-plane air gap. For the positive rod-rod air gap with one rod grounded, the average breakdown field strength is about 4.8 kV/cm, and it is about 5 kV/cm for negative rod-rod air gap. It can be seen that the breakdown field strength of an extremely nonuniform electric field decreases greatly compared with that of a uniform electric field, namely, about 30 kV/cm.

Under power frequency voltage, due to the polarity effect, the breakdown of a rod-plane gap occurs within the half cycle when the rod electrode is applied positive voltage and the voltage reaches the peak value. Test results showed that [2] when the gap distance $d < 250$ cm, the average breakdown field strength of rod plane air gap is about 4.8 kV/cm (peak value), and that of rod-rod air gap is about 5.36 kV/cm. In contrast, the breakdown voltage of rod-rod gap is higher than that of rod-plane gap, since the electric field distribution of the rod-rod gap is more uniform.

When $d > 40$ cm, the power frequency breakdown voltage (peak value) U_b of the rod-plane and rod-rod gap can be approximately estimated by the following formulas [3]:

$$\begin{array}{ll} \text{Rod-plane gap} & U_b = 40 + 5d \\ \text{Rod-rod gap} & U_b = 70 + 5.25d \end{array} \quad (5.5)$$

As the gap distance increases, the power frequency breakdown characteristic curve of long air gaps has a saturation phenomenon. However, the power frequency breakdown voltage required in power transmission and transformation projects is far from the serious saturation degree [4].

5.2 Breakdown Voltage Prediction of Typical Short Air Gaps

5.2.1 Power Frequency Breakdown Voltages of Typical Air Gaps

The sphere gap, rod-plane gap and rod-rod gap are three kinds of typical air gaps. IEEE Std 4-2013 [5] gives the experimental breakdown voltages of sphere gaps, while the power frequency voltage-withstand tests of rod-plane and rod-rod gaps are conducted to obtain their experimental breakdown voltages.

The experimental data of sphere gaps with the diameter D of 6.25, 10, 15 cm and gap distance d ranging from 1 to 5 cm are selected from IEEE Std 4-2013 and taken as sample data for breakdown voltage prediction study, as shown in Table 5.1.

For rod-plane and rod-rod gap, the rod electrode is made of brass with a hemispherical head, while the diameters (D_r) are 20, 25 and 30 mm, as shown in Fig. 5.3.

Table 5.1 Power frequency breakdown voltages of sphere gaps (peak value, kV)

d (cm)	*D* (cm)		
	6.25	10	15
1.0	31.9	31.7	31.7
1.5	45.5	45.5	45.5
2.0	58.5	59.0	59.0
2.6	72.0	74.5	75.5
3.0	79.5	84.0	85.0
3.5	87.5	95.0	98.0
4.0	95.5	105	110
4.5	101	115	122
5.0	107	123	133

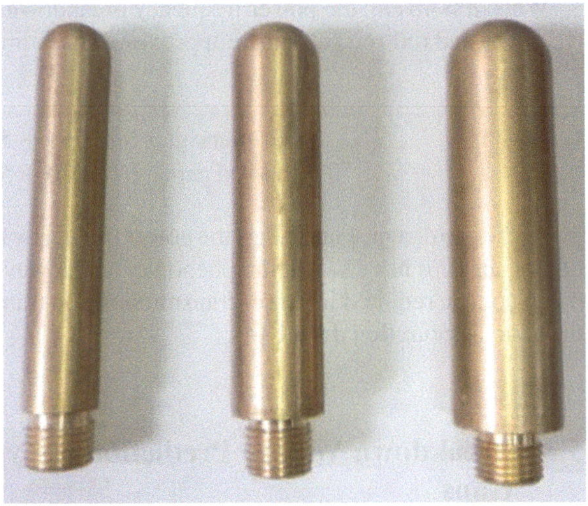

Fig. 5.3 Three kinds of rod electrodes with different diameters

Table 5.2 Power frequency breakdown voltages of rod-plane and rod-rod gaps (peak value, kV)

d (cm)	D_r (mm)					
	Rod-plane gap			Rod-rod gap		
	20	25	30	20	25	30
1.0	25.3	26.2	26.6	30.5	31.3	31.4
1.5	31.3	32.8	34.1	41.1	42.7	43.4
2.0	35.8	37.8	40.9	47.9	51.3	53.9
2.5	38.4	41.6	44.5	54.1	56.9	62.1
3.0	41.6	44.9	48.3	57.8	61.0	68.7
3.5	44.0	47.2	51.5	60.5	64.6	71.6
4.0	46.6	50.0	54.5	63.1	67.4	75.7
4.5	48.1	51.9	57.8	66.1	69.6	78.1
5.0	49.6	53.8	61.2	68.6	71.5	81.3

The plane electrode is a 2 mm thick copper plate with the size of 40 cm × 40 cm. Their power frequency breakdown voltages are measured by tests with the method of uniform increase voltage. The applied voltage is rose slowly and continuously until the gap breakdown. The breakdown voltage of each gap distance is measured for 5 times, and their average value is taken as the experimental result. The test data are corrected to standard atmospheric condition by *g* parameter method recommended in IEC 60060-1 [6]. The experimental breakdown voltages of rod-plane and rod-rod gaps are shown in Table 5.2.

In addition, the power frequency voltage-withstand tests are carried out for sphere-plane gap, rod-sphere gap, and sphere-sphere gap with different diameters. The test results are shown in Table 5.3, altogether 42 samples. For sphere-plane gaps, the sphere diameter is 10 cm. For small rod-sphere gaps, the small rod diameter is

Table 5.3 Power frequency breakdown voltages of sphere-plane gap, rod-sphere gap and sphere-sphere gap with different diameters (peak value, kV)

d (cm)	Sphere-plane	Small rod-sphere	Large rod-sphere	Small sphere-large sphere	Large sphere-small sphere
1.0	30.9	28.6	29.7	30.6	30.5
1.5	44.7	37.3	40.0	44.9	44.6
2.0	57.3	41.5	47.6	59.2	58.7
2.5	67.2	46.4	53.5	70.1	71.8
3.0	77.5	49.7	57.9	79.1	84.3
3.5	85.6	53.3	62.3	87.3	96.0
4.0	92.5	56.0	66.0	93.9	106.8
4.5	98.8	58.9	69.1	98.6	–
5.0	–	60.3	72.0	102.1	–

20 mm and the sphere diameter is 6 cm. For large rod-sphere gaps, the large rod diameter is 30 mm and the sphere diameter is 9.75 cm. For sphere-sphere gaps with different diameters, the large sphere diameter is 9.75 cm, and the small one is 6.5 cm.

5.2.2 *Analysis of the Electric Field Distributions*

The electric field distributions of sphere gaps, rod-plane and rod-rod gaps are calculated by FEM. For sphere gaps, the high voltage sphere is applied unit potential 1 V and the grounded sphere is applied zero potential. The electric field distributions along the shortest discharge path are shown in Fig. 5.4, with $D = 10$ cm, $d = 1–5$ cm, and $D = 6.25–15$ cm, $d = 3$ cm, respectively. For rod-plane gaps, the rod electrode is applied 1 V and the plane electrode is applied zero potential. The electric field distributions along the shortest discharge path are shown in Fig. 5.5, with $D_r = 20$ mm, $d = 1–5$ cm, and $D_r = 20–30$ mm, $d = 3$ cm. For rod-rod gaps, one rod is applied 1 V and the other is applied zero potential, the electric field distributions along the shortest discharge path are shown in Fig. 5.6.

The electric field distribution of the sphere gap is a U-shaped curve, with electric field concentration on the surface of both the high voltage and grounded sphere electrode. The electric field strength gradually decreases along the shortest path of the rod-plane gap, and the distribution is extremely nonuniform. The electric field distribution of the rod-rod gap is also a U-shaped curve, while the field strengths near two rod electrodes are much larger than those in the middle of the shortest path. The maximum field strengths of the three typical air gaps decrease with larger gap distances and high voltage electrode diameters.

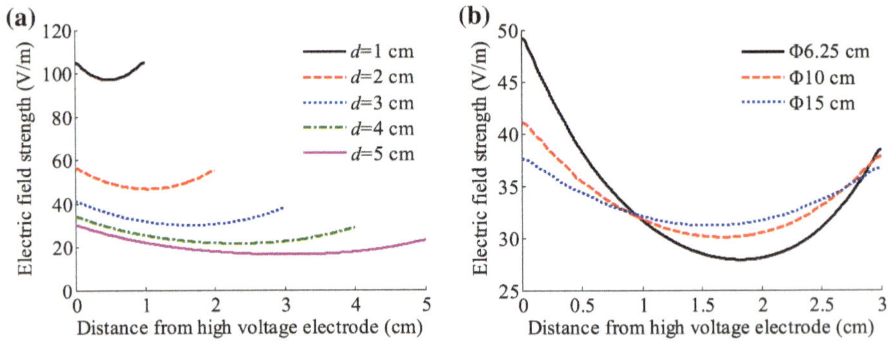

Fig. 5.4 Electric field distributions of sphere gaps along the shortest discharge path: **a** $D = 10$ cm, $d = 1$–5 cm, **b** $D = 6.25$–15 cm, $d = 3$ cm

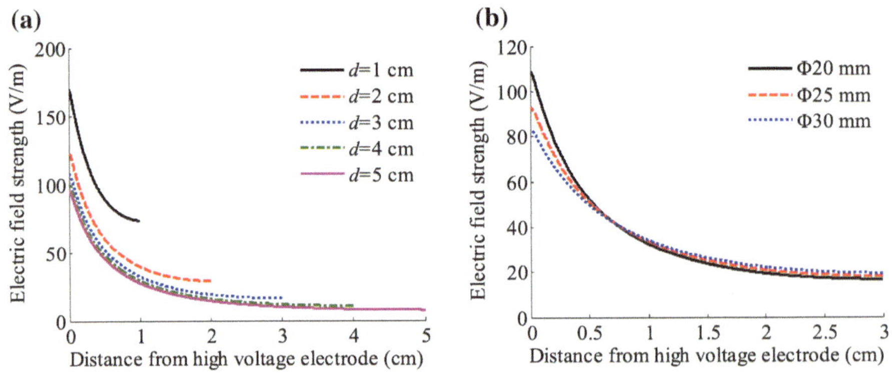

Fig. 5.5 Electric field distributions of rod-plane gaps along the shortest discharge path: **a** $D_r = 20$ mm, $d = 1$–5 cm, **b** $D_r = 20$–30 mm, $d = 3$ cm

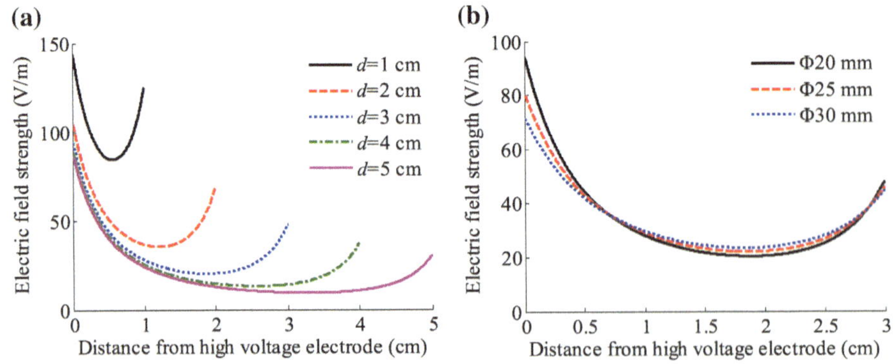

Fig. 5.6 Electric field distributions of rod-rod gaps along the shortest discharge path: **a** $D_r = 20$ mm, $d = 1$–5 cm, **b** $D_r = 20$–30 mm, $d = 3$ cm

The three typical air gaps respectively represent slightly uneven electric field, asymmetric and symmetric extremely nonuniform electric field. Taking the three gaps as training samples to train the SVM, it is helpful to improve the generalization performance of the prediction model.

5.2.3 *Orthogonal Design of Training Samples*

Orthogonal design is a scientific method to study the multi-factor optimization tests based on the orthogonal principle of the orthogonal array [7]. The orthogonal array is denoted as $L_a(b^c)$, where L represents the orthogonal table, a represents the number of tests, which is the number of rows in the table; c represents the number of factors, which is the number of columns in the table; b represents the levels of these factors. In order to reduce test work as much as possible, the orthogonal array is used to select training samples required for power frequency breakdown voltage prediction model of typical short air gaps. The basic idea is to select a number of test data from the three most common typical gaps, i.e., the sphere gap, rod-plane and rod-rod gap, to train the SVM model, thus to predict the breakdown voltages of air gaps with other gap distances or electrode structures.

The training samples are selected from Tables 5.1 and 5.2. According to the principle of orthogonal design, the factors and levels of typical short air gaps with different structures are listed in Table 5.4. The three levels of electrode tip size refer to different diameters of the high voltage electrode for the sphere gap, rod-plane and rod-rod gap. For sphere gaps, D_1, D_2 and D_3 respectively represent three diameters of 6.25 cm, 10 cm and 15 cm. For rod-plane and rod-rod gaps, D_1, D_2 and D_3 respectively represent three diameters of the hemispherical head, namely, 20 mm, 25 mm, and 30 mm. The three levels of gap distance d_1, d_2, d_3, represent different gap distances.

The first 3 columns in the $L_9(3^4)$ orthogonal array are used to construct the training sample set for breakdown voltage prediction of typical short air gaps with different structures, as shown in Table 5.5, altogether 9 samples. The other 72 data in Tables 5.1 and 5.2 are taken as test samples. The training sample set is used to train the SVM model, thus to predict the breakdown voltages of the test samples. In addition, in order to verify the generalization performance of the proposed method, the SVM model

Table 5.4 Factors and levels of typical air gaps with different structures

Levels	Factors		
	Gap type	Electrode tip size	Gap distance
1	Sphere gap	D_1	d_1
2	Rod-plane gap	D_2	d_2
3	Rod-rod gap	D_3	d_3

Table 5.5 Training sample set based on orthogonal design

No.	Factors			
	1-Gap type	2-Electrode tip size	3-Gap distance (cm)	U_b (kV)
1	Sphere gap	Φ6.25 cm sphere	1	31.9
2	Sphere gap	Φ10 cm sphere	3	84.0
3	Sphere gap	Φ15 cm sphere	5	133
4	Rod-plane gap	Φ20 mm rod	3	41.6
5	Rod-plane gap	Φ25 mm rod	5	53.8
6	Rod-plane gap	Φ30 mm rod	1	26.6
7	Rod-rod gap	Φ20 mm rod	5	68.6
8	Rod-rod gap	Φ25 mm rod	1	31.3
9	Rod-rod gap	Φ30 mm rod	3	68.7

is further used to predict the power frequency breakdown voltages of sphere-plane gaps, rod-sphere gaps, and sphere-sphere gaps with different diameters, as shown in Table 5.3.

5.2.4 *Prediction Results and Analysis*

The air gap breakdown voltage prediction model is established by LIBSVM toolbox, which was developed by Prof. Lin Chin-Jen, et al. in National Taiwan University [8]. This toolbox is easy to use, and it can effectively solve the classification problems involved in this book. The breakdown voltages of the 72 test samples of three typical short air gaps are firstly predicted and the influences of parameter optimization methods and feature dimension reduction methods on the breakdown voltage prediction results are analyzed.

5.2.4.1 Influence of Parameter Optimization Method

Under the sense of 3-CV, the improved GS algorithm, GA and PSO algorithm are respectively used to optimize the SVM parameters, so as to analyze their influences on the prediction results. For the improved GS algorithm, the optimal parameters are searched twice. In the first search, the search ranges of both the penalty factor C and the kernel parameter γ are $[2^{-10}, 2^{10}]$, and the step sizes are both 2^1. In the second search, the ranges of C and γ are respectively $[2^3, 2^9]$ and $[2^{-8}, 2^{-2}]$, while the step size is $2^{0.1}$. The parameter optimization results by the improved GS algorithm are shown in Fig. 5.7.

For parameter optimization by GA, the population quantity is set as 20, the maximum generation is 200, and the cross probability is 0.9. The search ranges of C and

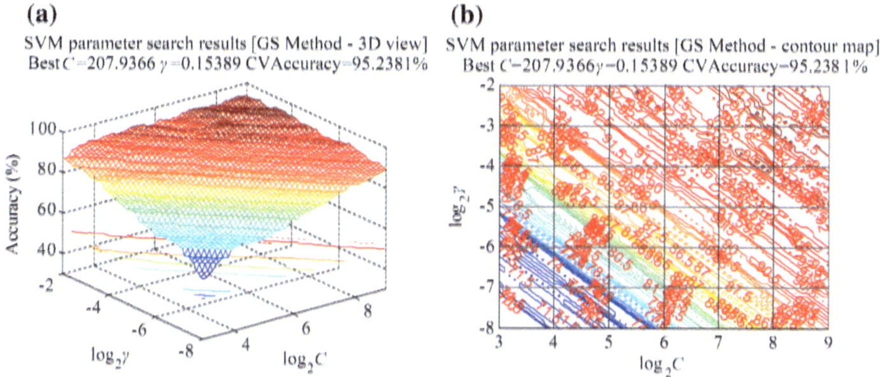

Fig. 5.7 Parameter optimization results by the improved GS algorithm: **a** 3D view, **b** contour map

Fig. 5.8 Parameter optimization results by GA

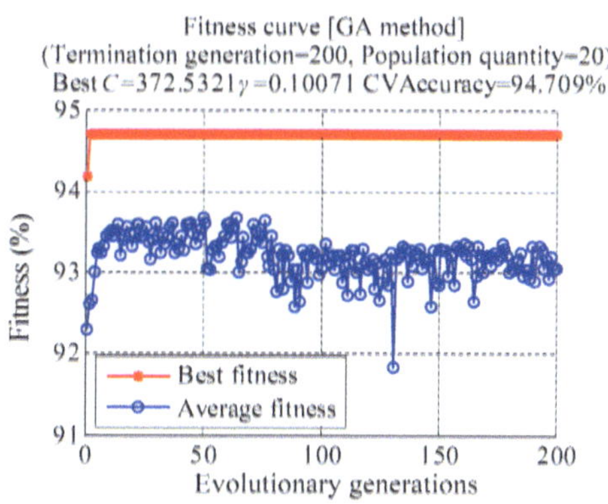

γ are respectively set as $[2^3, 2^9]$ and $[2^{-8}, 2^{-2}]$. For parameter optimization by PSO algorithm, the learning factors are set as $c_1 = 1.5$, $c_2 = 1.7$, the population quantity is 30, the maximum generation is 200, the coefficient $k = 0.6$, and the elastic coefficient $w = 1$. The search ranges of C and γ are $[2^3, 2^9]$ and $[2^{-8}, 2^{-2}]$. The fitness curves of GA and PSO algorithm are shown in Fig. 5.8 and Fig. 5.9, respectively.

The parameter optimization results by the three algorithms and the error indexes of their predicted results are summarized in Table 5.6. It can be seen that the PSO algorithm has the highest prediction accuracy, but the longest search time. The prediction accuracy of the improved GS algorithm is a bit lower than that of the PSO algorithm, but it spends much less search time. The search time of GA is the shortest, but its prediction error is the largest. Generally speaking, the error indexes of the predicted results obtained by the three optimization algorithms are very close, especially the results of the improved GS algorithm and the PSO algorithm are almost the same. Since the GA and the PSO are heuristic algorithms, the optimal parameters of different searches, using the same training samples, change every time. Therefore, the improved GS algorithm is finally selected.

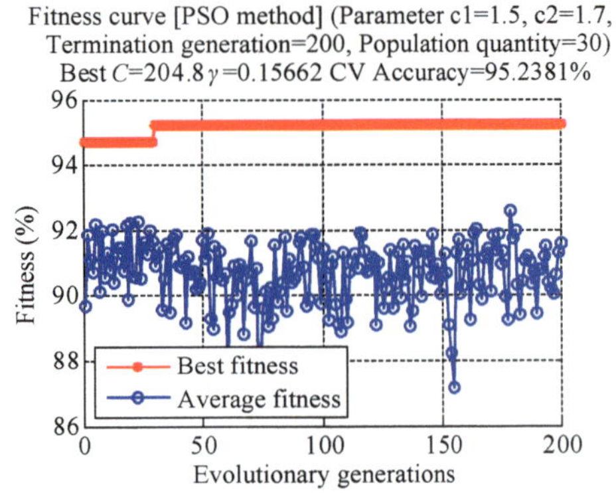

Fig. 5.9 Parameter optimization results by PSO

Table 5.6 Parameter optimization results and error indexes of the predicted results by different optimization algorithms

Results	Algorithms		
	Improved GS algorithm	GA	PSO algorithm
C	207.937	372.5320	204.8
γ	0.1539	0.1007	0.1566
t (s)	22.63	13.14	45.02
e_{SSE}	316.2166	351.2624	309.9437
e_{MSE}	0.2470	0.2603	0.2445
e_{MAPE}	0.0239	0.0246	0.0239
e_{MSPE}	0.0038	0.0039	0.0038

5.2.4.2 Influence of Electric Field Feature Dimension Reduction Methods

The correlation analysis method and PCA method are respectively applied to reduce the electric field feature dimensions, so as to analyze the influence of different input feature dimensions on the predicted results.

For the correlation analysis method, the correlation coefficient matrixes of two groups of electric field features, namely, the discharge channel and the shortest discharge path, are calculated respectively. Firstly, the features whose correlation coefficients with the breakdown voltage $r < 0.1$ are rejected, and the remained features are 21-d. Then the features whose correlation coefficients with the breakdown voltage $r < 0.3$ are further rejected, and remain 9-d. In addition, when the cross correlation coefficients of two or multiple features are larger than 0.9, the feature which has the strongest correlation with the breakdown voltage is retained, while the others are rejected, the remained features are 19-d.

For the PCA method, the principal components whose accumulated variance contribution rate is larger than 95% are selected from the two groups of electric field features. The number of the principal components after feature dimension reduction

Table 5.7 Optimal parameters and error indexes of the predicted results under different feature dimensions

Parameters and errors	Correlation analysis method				PCA method
	28	21	9	19	10
C	207.937	97.0059	512	238.856	157.586
γ	0.1539	0.2176	0.2333	0.0625	0.25
e_{SSE}	316.2166	305.6146	464.5894	372.0502	324.2234
e_{MSE}	0.2470	0.2428	0.2994	0.2679	0.2501
e_{MAPE}	0.0239	0.0283	0.0350	0.0297	0.0250
e_{MSPE}	0.0038	0.0041	0.0051	0.0046	0.0039

is 10-d. The optimal parameters and error indexes of the prediction model under different feature dimensions are shown in Table 5.7.

For feature dimension reduction by the correlation analysis method, it can be seen from Table 5.7 that the e_{SSE} and e_{MSE} under 21-d decrease slightly, e_{MAPE} and e_{MSPE} increase slightly compared with those under 28-d. The errors under 9-d are the largest, which indicates that too less input features will reduce the prediction accuracy of the SVM model. The error indexes under 19-d are also larger than 28-d. For feature dimension reduction by the PCA method, the error indexes under 10 principal components are a bit larger than those under 28-d. In general, the feature dimension reduction has little influence on the breakdown voltage prediction results. Therefore, the 28-d electric field features are finally applied.

5.2.4.3 Breakdown Voltage Prediction Results

The predicted results of the 72 test samples, including the sphere gaps, rod-plane and rod-rod gaps, are shown in Figs. 5.10 and 5.11, while the SVM model is optimized by the improved GS algorithm under 28-d electric field features. For sphere gaps, the predicted results of those gaps with the sphere diameter of 10 and 15 cm are taken for example and plotted in Fig. 5.10. In Figs. 5.10 and 5.11, the data points labelled by triangle, circular and square markers are experimental values, and the linear tags represent the predicted values. It can be seen that except for few samples, the breakdown voltage prediction results of other test samples are in good agreement with the experimental data. The maximum relative error of the predicted results is 9%, and the e_{MAPE} is 2.39%. The results indicate that the SVM model trained by orthogonal samples has high accuracy to predict the power frequency breakdown voltages of typical short air gaps [9].

In order to further verify the generalization performance of the SVM model, the breakdown voltages of the sphere-plane gaps, rod-sphere gaps, and sphere-sphere gaps with different diameters shown in Table 5.3 are predicted respectively. The

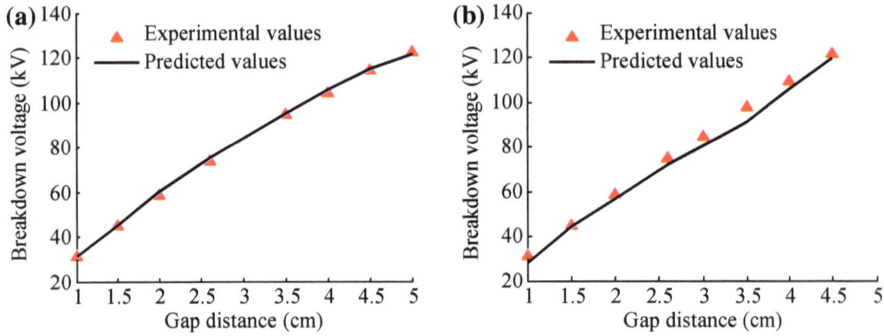

Fig. 5.10 Comparison between predicted and experimental values of sphere gap breakdown voltage: **a** $D = 10$ cm, **b** $D = 15$ cm

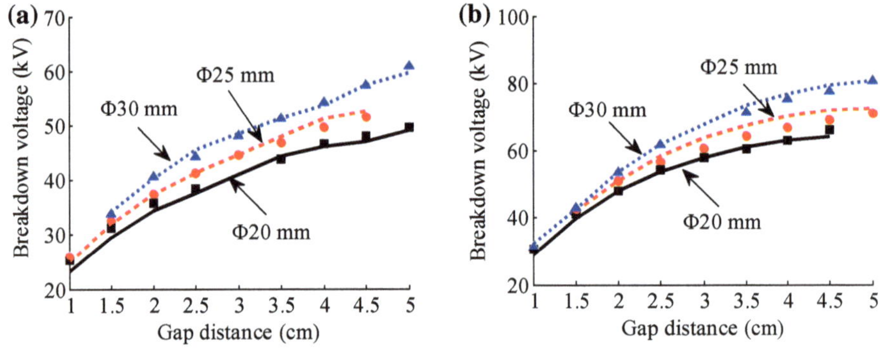

Fig. 5.11 Comparison between predicted and experimental values of rod-plane and rod-rod gap breakdown voltage: **a** rod-plane gap, **b** rod-rod gap

results are shown in Table 5.8. It can be seen that the errors indexes of the 5 kinds of gaps are within reasonable range, while the e_{MAPE}s are respectively 1.75%, 5.22%, 2.78%, 3.22% and 4.57%.

The comparisons between the predicted and experimental breakdown voltages of the above-mentioned short air gaps are shown in Fig. 5.12. It can be seen that the predicted breakdown voltages of most samples agree well with the experimental data. This conclusion indicates that the SVM model trained by the experimental data of sphere gaps, rod-plane gaps and rod-rod gaps has good prediction accuracy for the sphere-plane gaps, rod-sphere gaps and sphere-sphere gaps with different diameters. In general, the SVM model trained by 9 orthogonal samples is applicable to accurately predict the breakdown voltages of the 72 test samples, i.e., the three typical gap types and the 42 test samples, i.e., the other three gap types consist of typical electrodes [9]. The results verify the superiority of the SVM model in dealing with small sample problem, and also indicate that the proposed method is of certain generalization performance.

Table 5.8 Optimal parameters and error indexes of breakdown voltage prediction results of sphere-plane gaps, rod-sphere gaps and sphere-sphere gaps with different diameters

Parameters and errors	Sphere-plane	Small rod-sphere	Large rod-sphere	Small sphere-large sphere	Large sphere-small sphere
C	207.937	137.187	207.937	512	222.861
γ	0.125	0.1539	0.2333	0.2031	0.125
e_{SSE}	9.7732	73.5057	53.7090	58.8873	75.3146
e_{MSE}	0.3908	0.9526	0.8143	0.8526	1.2398
e_{MAPE}	0.0175	0.0522	0.0278	0.0322	0.0457
e_{MSPE}	0.0100	0.0191	0.0121	0.0169	0.0202

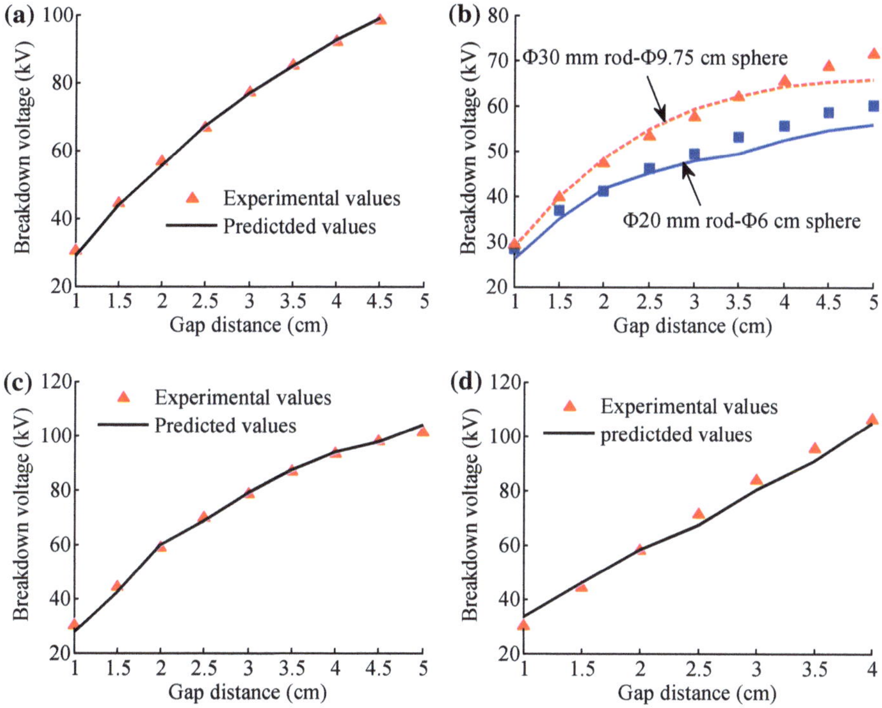

Fig. 5.12 Comparisons between predicted experimental breakdown voltages of sphere-plane gaps, rod-sphere gaps and sphere-sphere gaps with different diameters: **a** Φ10 cm sphere-plane gap, **b** rod-sphere gap, **c** Φ6.5 cm sphere-Φ9.75 cm sphere gap, **d** Φ9.75 cm sphere-Φ6.5 cm sphere gap

5.3 Breakdown Voltage Prediction of Atypical Short Air Gaps

In order to further verify the applicability of the proposed method to breakdown voltage prediction of atypical air gaps, the SVM model trained by the experimental data of typical air gaps, like sphere gaps and rod-plane gaps, is used to predict the power frequency breakdown voltages of serial gaps, ring and stranded conductor air gaps.

5.3.1 Power Frequency Breakdown Voltages of Atypical Air Gaps

The serial gaps include sphere-plane-sphere, rod-plane-sphere and rod-plane-rod gaps, their power frequency voltage-withstand tests are carried out respectively. The experimental arrangement diagram is shown in Fig. 5.13. The diameters of the large sphere and the small sphere electrode are respectively 9.75 cm and 6.5 cm. The rod electrode is a hemispherical rod whose diameter is 3 cm. The plane electrode is a 3 mm thick square plate with a smooth surface, whose side length is 30 cm. All of the electrodes are made of brass. The high voltage and grounded electrodes are fixed on an insulating support, and the plane electrode is suspended in the middle of the gap, supported by a post insulator. Taking rod-plane-sphere gap for example, the discharge channel and the shortest discharge path of the serial gaps are also shown in Fig. 5.13.

For the ring gaps, two grading rings used on transmission line are taken as the electrodes, and therefore constitute two ring-plane gaps and a ring-ring gap. The ring electrodes are made of aluminium alloy. The ring diameter and pipe diameter of the large ring electrode are 26.4 and 3.2 cm, and these two diameters are 26 and 2 cm for the small ring electrode. The plane electrode is the same with that used in serial gaps. The experimental arrangement diagram of ring gap is shown in Fig. 5.14. For the ring-plane gaps, the ring electrode is applied high voltage and the plane electrode is grounded. For the ring-ring gap, the small ring is applied high voltage and the large ring is grounded.

Taking the ring-plane gap for example, the discharge channel and the shortest discharge path are shown in Fig. 5.14 as well. For the ring-plane gap, the discharge channel is the spatial projective region from the ring electrode to the plane electrode along the horizontal direction. For the ring-ring gap, the discharge channel is the spatial projective region from the small ring electrode to the large ring electrode along the horizontal direction. The shortest discharge path means the shortest geometric path from the high voltage electrode to the grounded electrode.

In addition, the power frequency breakdown voltages of stranded conductor-plane and stranded conductor-sphere gaps were measured in [10]. The structure diagram of stranded conductor air gaps is shown in Fig. 5.15. The conductor electrode is an

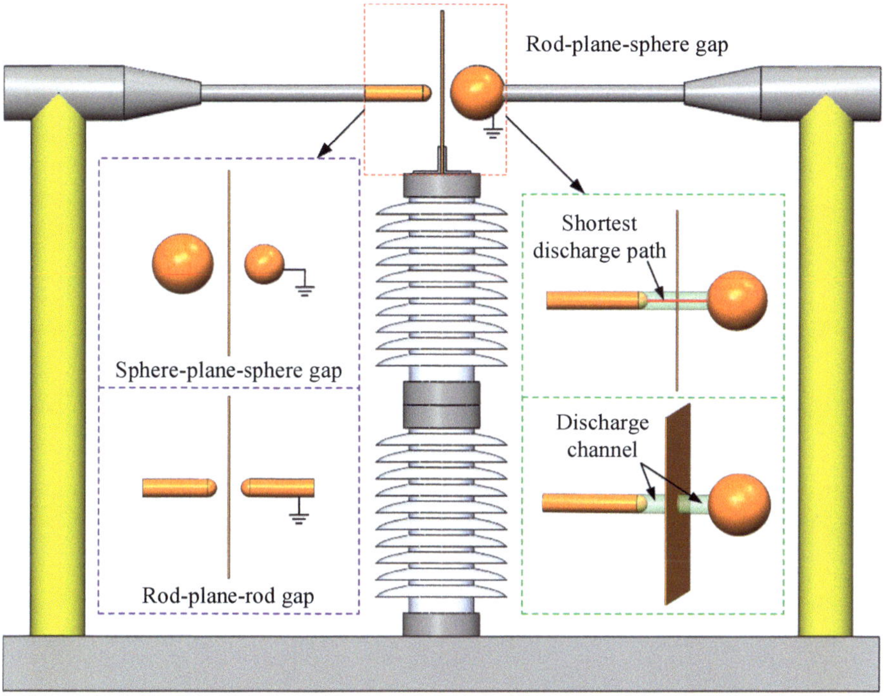

Fig. 5.13 Experimental arrangement diagram of serial gaps and definitions of the discharge channel and the shortest discharge path

all-aluminum, 7-strand, 39.3 mm^2, about 3 m long conductor. The plane electrode is a 2.4 m $\times$ 3 m grounded aluminum sheet, and the sphere electrode is also made of aluminum, while the diameter is 10.16 cm. The stranded conductor is floating above the plane or the sphere electrode, and it is applied high voltage, while the plane and the sphere electrode are grounded. The definitions of the discharge channel and the shortest discharge path for the stranded conductor air gaps are also shown in Fig. 5.15.

The experimental power frequency breakdown voltages of the 3 kinds of atypical short air gaps are shown in Table 5.9. For serial air gaps, the gap spacing d is the shortest distance between the high voltage electrode and the grounded electrode, without considering the decrease of the distance caused by the thickness of the plane electrode.

For serial air gaps, d refers to the shortest distance between the high voltage electrode and the low voltage electrode, not considering the spacing decreasing caused by the thickness of the plane. The above gaps are taken as test samples, and their breakdown voltages are predicted by the SVM model trained by experimental data of typical air gaps. The predicted results are compared with the experimental data to validate the effectiveness of the prediction method.

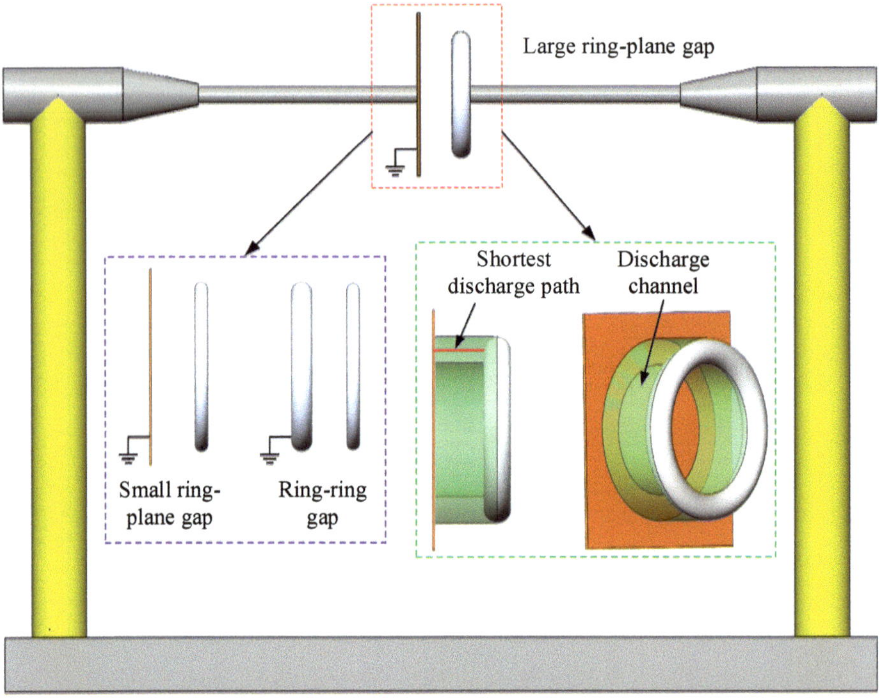

Fig. 5.14 Experimental arrangement diagram of ring gaps and definitions of the discharge channel and the shortest discharge path

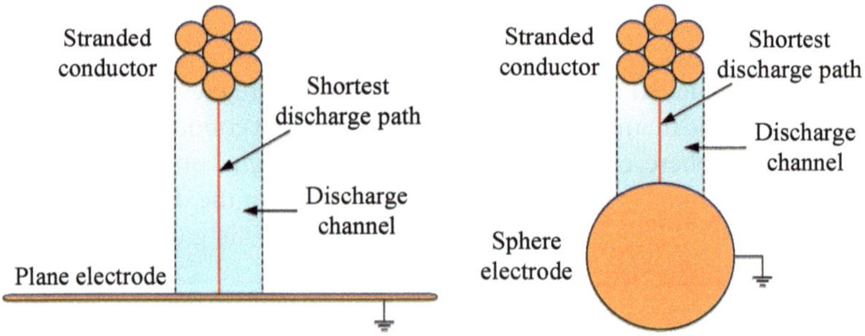

Fig. 5.15 Structure diagram of stranded conductor air gaps

5.3.2 Analysis of the Electric Field Distributions

The FEM models of the above three types of atypical air gaps are built according to the experimental arrangements. Applying 1 V on the high voltage electrode, and 0 V on the grounded electrode, the electric field distribution of each gap distance is calculated and the electric field features are extracted after post-processing of the

Table 5.9 Experimental power frequency breakdown voltages of atypical short air gaps (peak values)

d (cm)	Breakdown voltage (kV)							
	Sphere-plane-sphere	Rod-plane-sphere	Rod-plane-rod	Large ring-plane	Small ring-plane	Ring-ring	Stranded conductor-plane	Stranded conductor-sphere
1	–	–	–	28.2	25.5	26.8	–	–
2	44.4	33.8	36.9	50.6	47.4	49.7	34.5	40.5
3	63.0	46.9	49.0	68.2	62.2	67.9	41.1	50.6
4	78.5	54.6	58.4	85.2	72.5	82.7	45.9	57.2
5	93.0	62.0	62.0	98.9	81.0	96.4	50.7	65.0
6	107.6	70.0	68.1	110.8	88.2	106.6	53.3	69.2

calculation results. It should be noted that the plane electrode in the middle of the serial air gap is set as floating potential.

Taking the rod-plane-sphere gap, the ring-ring gap and the stranded conductor-plane gap as examples, when $d = 4$ cm, the cloud charts of their electric field distributions are shown in Fig. 5.16. The electric field distributions of the atypical air gaps along the shortest discharge path are shown in Fig. 5.17. For the serial gaps, the thickness of the plane electrode is ignored when plotting the electric field distribution curve.

It can be seen from Figs. 5.16a and 5.17a that the maximum electric field strength of the serial gap appears on the surface of the high voltage electrode, gradually decreases along the shortest path to the plane electrode, and then gradually increases from the plane to the grounded electrode. The electric field distributions of the two serial gaps on two sides of the plane electrode are similar to those of typical gaps, such as rod-plane gap. The electric field distribution of sphere-plane-sphere gap is more uniform than that of rod-plane-sphere and rod-plane-rod gap, while the latter two are almost the same, with only a little difference near the grounded electrode. Therefore, it can be deduced that the breakdown voltages of rod-plane-sphere and rod-plane-rod gap are nearly equal under the same gap distance, while that of sphere-plane-sphere gap is the largest. The experimental data shown in Table 5.9 are basically in line with this law.

It can be seen from Fig. 5.17b that the electric field distributions of ring-plane gaps are similar to that of rod-plane gap, while the field strength gradually decreases from the ring electrode to the grounded plane electrode. For the ring-ring gap, the electric field distribution along the shortest discharge path is a U-shape curve, which has similarity with the sphere gap. On the other hand, there is little difference between the electric field magnitude of large ring-plane gap and ring-ring gap, both of them are less than that of small ring-plane gap. Hence, it can be deduced that the breakdown voltages of large ring-plane gap and ring-ring gap are almost the same, while that of the small ring-plane gap is much smaller. The analysis results coincide with the experimental data shown in Table 5.9.

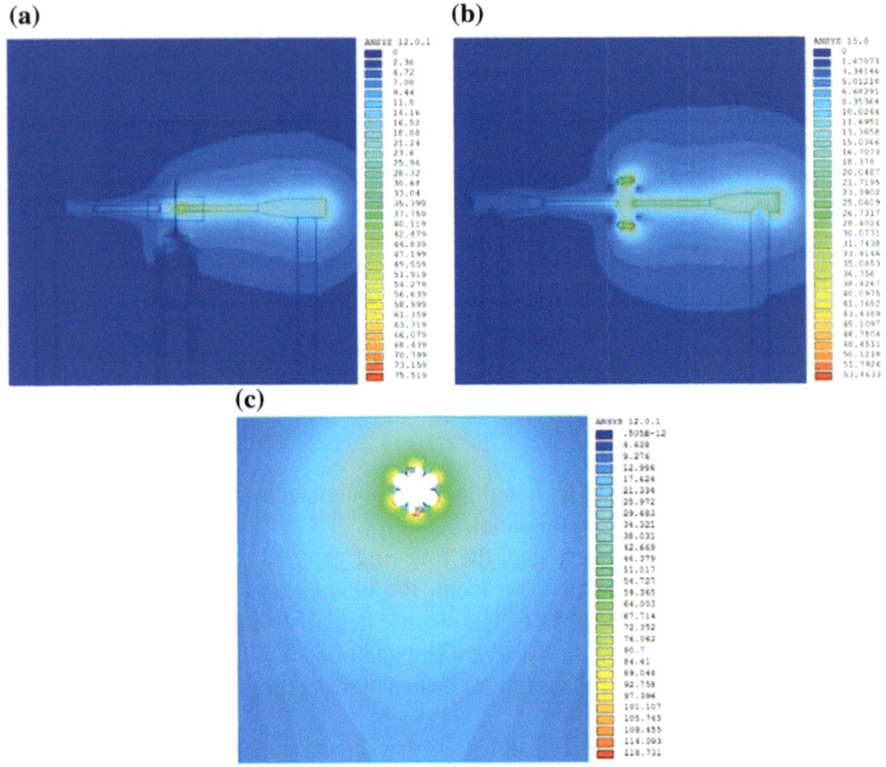

Fig. 5.16 Electric field distributions of atypical air gaps (d = 4 cm): **a** rod-plane-sphere gap, **b** ring-ring gap, **c** stranded conductor-plane gap

Similarly, as shown in Fig. 5.17c, the electric field distribution of stranded conductor-plane gap is similar to that of rod-plane gap, and the stranded conductor-sphere gap has some similarity with the sphere gap. Besides, the electric field distributions agree with the experimental results shown in Table 5.9.

5.3.3 *Training Sample Set*

According to the above-mentioned analysis results, the electric field distributions along the shortest discharge paths of the investigated atypical air gaps are similar with either sphere gap or rod-plane gap. Therefore, several experimental data of the breakdown voltages of sphere gaps and rod-plane gaps are selected as the training sample set for the SVM model used to predict the breakdown voltages of the atypical air gaps. The training sample set is shown in Table 5.10, including 3 data of sphere gaps and also 3 data of rod-plane gaps, with different diameters of sphere or rod

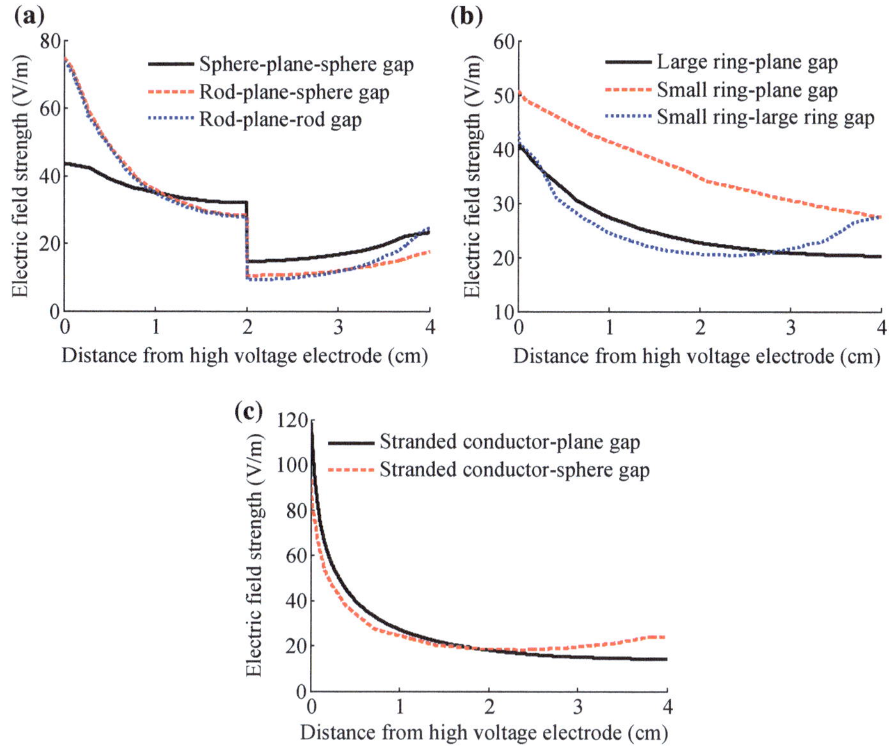

Fig. 5.17 Electric field distributions of the atypical air gaps along the shortest discharge path: **a** serial air gaps, **b** ring structure air gaps, **c** stranded conductor air gaps

Table 5.10 Training sample set and selection method

Training sample set				Training sample selection method
Gap type	D (cm)	d (cm)	U_b (kV)	
Sphere gap	6.25	1	31.9	Sphere-plane-sphere: S Rod-plane-sphere: S + R Rod-plane-rod: R Ring-plane: R Ring-ring: S Conductor-plane: R Conductor-sphere: S
	10	3	84.0	
	15	5	133	
Rod-plane gap	2.0	3	41.6	
	2.5	5	53.8	
	3.0	1	26.6	

electrode and different gap distances. D is the diameter of the sphere or rod electrode, d is the gap distance and U_b is the crest value of power frequency breakdown voltage.

In Table 5.10, the sphere gap and rod-plane gap are respectively labeled as "S" and "R". For the breakdown voltage prediction of sphere-plane-sphere gaps, ring-ring gaps and stranded conductor-sphere gaps, the training samples used to train the SVM model are the 3 experimental data of sphere gaps (S). For rod-plane-rod gaps,

Table 5.11 Error indexes of the breakdown voltage prediction results of atypical short air gaps

Error indexes	Sphere-plane-sphere	Rod-plane-sphere	Rod-plane-rod	Ring-plane	Ring-ring	Stranded conductor-plane	Stranded conductor-sphere
e_{SSE}	12.8766	14.7029	42.3133	107.0189	10.4520	7.8604	36.7387
e_{MSE}	0.7177	0.7669	1.3010	0.8621	0.5388	0.5607	1.2122
e_{MAPE}	0.01	0.03	0.054	0.0367	0.02	0.022	0.05
e_{MSPE}	0.0072	0.0143	0.0275	0.0135	0.0097	0.0134	0.0252

ring-plane gaps and stranded conductor-plane gaps, the training samples are the 3 experimental data of rod-plane gaps (R). For rod-plane-sphere gaps, the training samples are the 6 experimental data of sphere gaps and rod-plane gaps (S + R).

5.3.4 Prediction Results and Analysis

The training samples shown in Table 5.10 are used to train the SVM model. The input parameters are the electric field features, some after feature dimension by correlation analysis method. The improved GS algorithm is used for parameter optimization of the SVM model. For different gap configurations, the error indexes of the predicted results are shown in Table 5.11.

It can be seen that the e_{MAPE}s of the 3 kinds of serial gaps are respectively 1%, 3% and 5.4%. For ring-plane and ring-ring gaps, the e_{MAPE}s are 3.67% and 2%, and for stranded conductor-plane and stranded conductor-sphere gaps, they are 2.2% and 5% respectively. Comparisons between the predicted results and experimental values of serial gaps, ring gaps and stranded conductor gaps are respectively shown in Fig. 5.18, Fig. 5.19 and Fig. 5.20. It can be seen that these predicted values are in good accordance with experimental values, which validates the validity of the proposed method and SVM model for breakdown voltage prediction of atypical air gaps.

It can be seen from the above predicted results that [11] the SVM model trained by a few experimental data of typical air gaps, which is a small sample set, is applicable to predict the breakdown voltages of those atypical gap geometries with the similar electric field distributions.

5.4 Brief Summary

In this chapter, the power frequency breakdown voltages of typical air gaps and atypical air gaps are predicted by the proposed air insulation prediction model. The influences of the parameter optimization methods and feature dimensions on the prediction results are analyzed. Some conclusions can be drawn as follows:

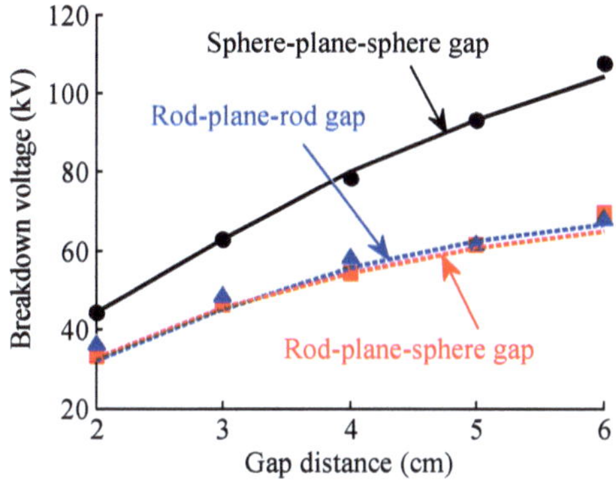

Fig. 5.18 Comparison between the predicted and experimental breakdown voltages of serial air gaps

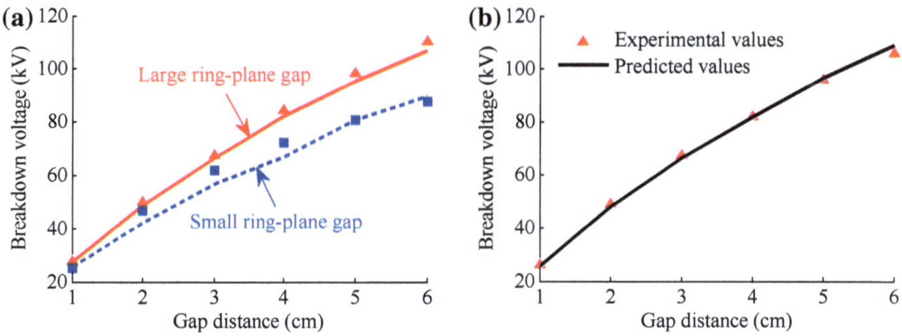

Fig. 5.19 Comparison between the predicted and experimental breakdown voltages of ring structure air gaps: **a** ring-plane gap, **b** ring-ring gap

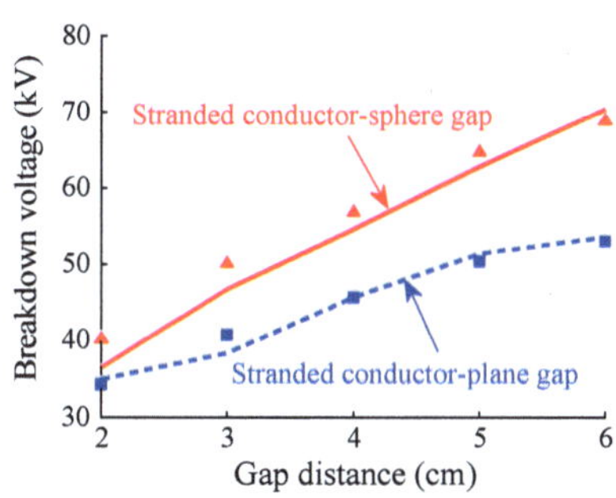

Fig. 5.20 Comparison between the predicted and experimental breakdown voltages of stranded conductor air gaps

1. For parameter optimization of the SVM model, under the sense of 3-CV, the prediction accuracy of the model optimized by 3 algorithms can be sorted as PSO algorithm > improved GS algorithm > GA. The results of improved GS algorithm and PSO algorithm have little difference. As for the parameter search time, that is PSO algorithm > improved GS algorithm > GA. In general, the error indexes of the SVM prediction results optimized by 3 algorithms are very close. Since the parameter search results of GA and PSO algorithm are unstable, the optimization algorithms should be selected according to the actual application situation.
2. The correlation analysis method and PCA method are used to reduce the electric field feature dimension. The prediction accuracy after feature dimension reduction is lower than that under all 28-d features, but only with little decline. In general, the feature dimension reduction does not have significant influence on the predicted results.
3. With 28-d input electric field features, the SVM model optimized by the improved GS algorithm and trained by 9 orthogonal samples is applied to predict the power frequency breakdown voltages of sphere gaps, rod-plane gaps and rod-rod gaps. The predicted values are in good agreement with the experimental data and the e_{MAPE} of 72 test samples is only 2.39%.
4. The power frequency breakdown voltages of sphere-plane gaps, rod-sphere gaps and sphere-sphere gaps with different diameters are predicted by the SVM model. The e_{MAPE}s of the 5 groups of test samples are respectively 1.75%, 5.22%, 2.78%, 3.22% and 4.57%. The predicted results verify the generalization performance of the prediction method to be applied to different electrode structures.
5. For serial air gaps including the sphere-plane-sphere gap, rod-plane-sphere gap and rod-plane-rod gap, the electric field distributions on two sides of the plane electrode are similar to those of typical air gaps. The SVM model trained by the experimental data of typical air gaps like sphere gaps and rod-plane gaps are used to predict the breakdown voltages of serial gaps. The predicted results coincide well with the experimental data. For the 3 kinds of serial gaps, the e_{MAPE}s of the breakdown voltage prediction results are 1%, 3% and 5.4%, respectively.
6. The electric field distributions of ring structure air gaps and stranded conductor air gaps are also similar to those of typical air gaps, and their breakdown voltages can also be predicted by the SVM model trained by the experimental data of typical air gaps. For the ring-plane and ring-ring gaps, the e_{MAPE}s of the predicted results are respectively 3.67% and 2%. For stranded conductor-plane gaps and stranded conductor-sphere gaps, the e_{MAPE}s are 2.78% and 4.67% respectively. The results verify the validity of the proposed method for breakdown voltage prediction of atypical short air gaps.

References

1. IEC 60052 (2002) Voltage measurement by means of standard air gaps
2. Yan Z, Zhu DH (2007) High voltage insulation technology, 2nd edn. China Electric Power Press, Beijing
3. Guan GZ (2003) Fundamentals of high voltage engineering. China Electric Power Press, Beijing
4. Wan QF, Huo F, Xie L et al (2012) Summary of research on flashover characteristics of long air-gaps. High Volt Eng 38(10):2499–2505
5. IEEE Std 4 (2013) IEEE standard for high-voltage testing techniques
6. IEC 60060-1 (2011) High-voltage test techniques—part 1: general definitions and test requirements
7. JB/T 7510 (1994) Optimization method of technological parameters—orthogonal experiment
8. Chang CC, Lin CJ (2011) LIBSVM: a library for support vector machines. ACM Trans Intell Syst Technol 2(3):1–27
9. Qiu ZB, Ruan JJ, Huang DC et al (2016) Hybrid prediction of the power frequency breakdown voltage of short air gaps based on orthogonal design and support vector machine. IEEE Trans Dielectr Electr Insul 23(2):795–805
10. Barsch JA, Sebo SA, Kolcio N (1999) Power frequency AC sparkover voltage measurements of small air gaps. IEEE Trans Power Deliv 14(3):1096–1101
11. Qiu ZB, Ruan JJ, Huang CP et al (2016) A method for breakdown voltage prediction of short air gaps with atypical electrodes. IEEE Trans Dielectr Electr Insul 23(5):2685–2694

Chapter 6
Impulse Discharge Voltage Prediction of Air Gaps

6.1 Air Gap Breakdown Characteristics Under Impulse Voltages

6.1.1 Lightning Impulse Breakdown Characteristics

Lightning wave is an aperiodic pulse, whose waveform parameters are of statistical property. The standard lightning impulse voltage waveform specified in IEC standard is shown in Fig. 2.3b. Its virtual front time $T_f = 1.2$ μs and the allowable deviation is ±30%. The time to half-value $T_2 = 50$ μs, and the allowable deviation is ±20%.

The duration time of the lightning impulse voltage is very short, which is comparable with the time required for air gap breakdown. Therefore, the gap breakdown characteristics are affected by the voltage action time. When an air gap is applied the impulse voltage, the gap breakdown requires not only enough voltage amplitude, but also a certain duration time which makes the discharge develop to breakdown.

As shown in Fig. 6.1, after the time of t_0, the voltage rises from zero to the static breakdown voltage U_0, but the gap is not breakdown at this moment. Only when an electron appears in the gap and causes ionization process, the discharge is possible to develop and eventually leads to breakdown. This electron is called the effective electron. The appearance of effective electrons is random and takes a certain period of time. The required time from t_0 to the appearance of an effective electron in the gap is called the statistical delay t_s. From the appearance of an effective electron to the formation of electron avalanche until the gap breakdown completely, it also requires a certain period of time for discharge development, which is called discharge formation delay t_f. Hence, the whole discharge time t_b consists of three parts, that is

$$t_b = t_0 + t_s + t_f \tag{6.1}$$

where t_0 is voltage rise time and the sum of t_s and t_f is the discharge delay t_{lag}.

Z. Qiu et al., *Air Insulation Prediction Theory and Applications*, Power Systems,
https://doi.org/10.1007/978-981-10-5163-0_6

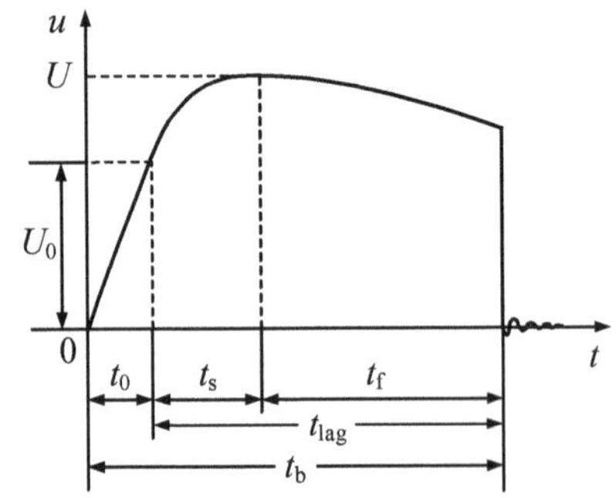

Fig. 6.1 Components of air gap discharge time under the impulse voltage

The discharge delay is related to the applied voltage and the uniformity of the electric field distribution. The required time for discharge is shorter when the gap is applied a greater voltage. In a short gap (within a few centimeters), especially when the electric field is uniform, the discharge formation delay t_f is short, and therefore the discharge delay is mainly determined by t_s. In a longer gap with nonuniform electric field and high local field strength, the probability for appearance of an effective electron increases, thus the discharge delay is mainly determined by t_f, and the more nonuniform the electric field is, the longer t_f is.

Since the discharge delay is of dispersion, it is difficult to determine the exact value of air gap discharge voltage under the impulse voltage. When the impulse voltage with the same waveform and peak value is applied on the gap for several times, sometimes the gap will breakdown and sometimes it will not. The voltage under which the breakdown probability is equal to 50% (50% impulse breakdown voltage U_{50}) is used in engineering to reflect the characteristics of air gap withstanding impulse voltage. The dispersion of the breakdown voltage is described by the standard deviation σ.

When U_{50} is used to determine the gap length in engineering, there should be a margin according to the dispersion of the breakdown voltage. In a uniform or slightly non-uniform electric field, the dispersion of the impulse breakdown voltage is very small, so there is little difference between U_{50} and the static breakdown voltage U_s. The impulse coefficient β, which is the ratio of U_{50} and U_s, is approximately equal to 1. Due to the short discharge delay, the breakdown under U_{50} usually occurs at the time near the peak value of the wavefront. In extremely nonuniform electric field, due to the long discharge delay, the impulse coefficient $\beta > 1$, and the breakdown voltage is of large dispersion, while the standard deviation can be $\pm 3\%$. The breakdown under U_{50} usually occurs at the wave tail.

The experimental 50% breakdown voltages of the standard rod-rod gaps under 1.2/5 and 1.2/50 μs lightning impulse voltages were given in IEEE Std4-1995. The rod-rod gap consists of two 12.5 mm square rod electrodes arranged horizontally, and the length of the rod electrode is greater than 1 m. The relationship between the U_{50} and the gap distance d is shown in Fig. 6.2. It can be seen that the lightning impulse breakdown voltages of the rod-rod gaps have an approximately linear relation

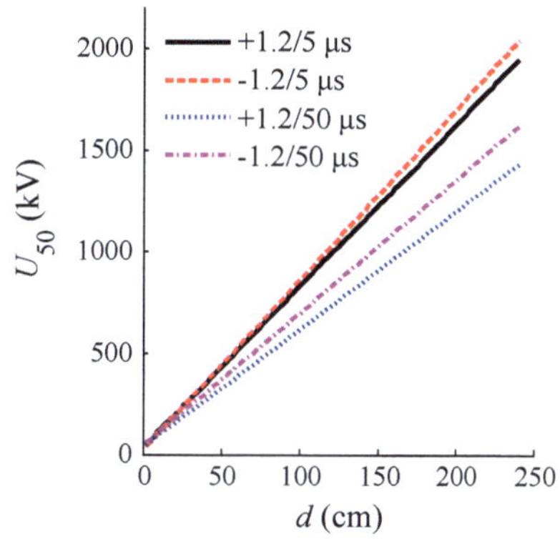

Fig. 6.2 Relationship between lightning impulse U_{50} of rod-rod gaps and the gap distance

with the gap distance. Due to the polarity effect, the positive lightning impulse breakdown voltage is less than the negative under the same gap distance, and their differences gradually increase with the longer gap distance. In addition, the U_{50} under the standard 1.2/50 μs lightning impulse is lower than that under the 1.2/5 μs short wave tail lightning impulse.

When the gap distance $d > 40$ cm, under 1.5/40 μs lightning impulse voltage, the U_{50} of rod-plane gap and rod-rod gap can be estimated by the following approximate formulas respectively [1]:

$$\text{Rod-plane gap}\begin{cases}\text{Positive polarity } U_{50} = 40 + 5d \\ \text{Negative polarity } U_{50} = 215 + 6.7d\end{cases}$$
$$\text{Rod-rod gap}\begin{cases}\text{Positive polarity } U_{50} = 75 + 5.56d \\ \text{Negative polarity } U_{50} = 110 + 6d\end{cases} \tag{6.2}$$

Since the duration time of lightning impulse voltage is very short, the discharge delay cannot be ignored. As a result, the air gap breakdown characteristics under impulse voltages are related to not only the 50% breakdown voltage, but also the discharge time. In engineering, the relation between the maximum voltage appeared during the breakdown process and the discharge time is used to characterize air breakdown characteristics under impulse voltages, which is called the volt-second or volt-time (U-t) characteristic. The voltage-time curve used to describe the relationship between the breakdown voltage and the discharge time is called volt-second or volt-time (U-t) characteristic curve.

The drawing method of U-t characteristic curve is shown in Fig. 6.3. Keep the impulse voltage waveform constant and increase the voltage peak. Under a low voltage, the gap will breakdown during the wave tail, and the crest voltage U_1 is taken as the gap breakdown voltage. The intersection point of U_1 and the discharge time t_1 is taken as a point of the U-t characteristic curve. Under a high voltage, gap breakdown occurs during the wavefront. The voltage at the breakdown time U_3 is

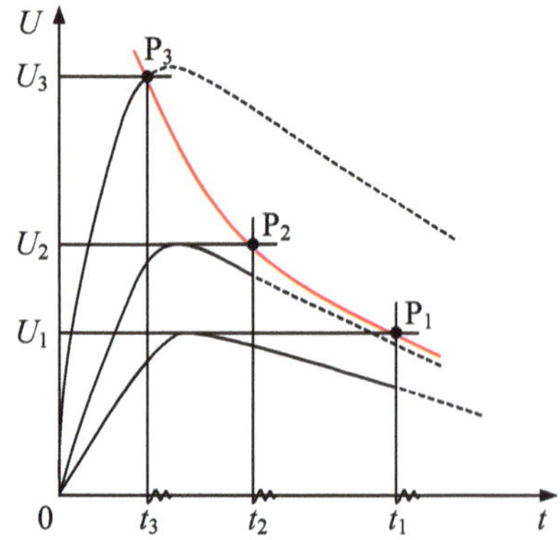

Fig. 6.3 Drawing method of *U-t* characteristic curve

taken as the gap breakdown voltage, and its intersection point with the discharge time t_3 is taken as a point of the *U-t* characteristic curve. By this way, a series of *U-t* intersections P_1, P_2, P_3 … can be obtained. The curve connecting these points is the *U-t* characteristic curve.

Since the discharge time is of dispersion, a series of discharge time can be obtained at each voltage level, therefore, the *U-t* characteristic is a band region bounded by the upper and lower envelope curve. In engineering, the air gap impulse breakdown characteristics are characterized by the average or 50% *U-t* characteristic curve, which is drawn by connecting each point of the average discharge time.

The shape of the *U-t* characteristic curve is related to the electric field distribution in the gap. For a uniform or slightly uneven electric field, the *U-t* curve is relatively flat. This is because the average field strength is high when breakdown occurs, with fast discharge development and short discharge delay. The *U-t* curve of the extremely nonuniform electric field is steep, because the average field strength is low when breakdown occurs, and the streamer always develops from high field area to weak field area, and the discharge velocity is affected by the electric field distribution, which prolongs the discharge delay.

The *U-t* characteristic plays an important role in insulation coordination. For example, the *U-t* characteristic curve of the equipment used for overvoltage protection (such as the arrester) should be as flat as possible, and the upper envelope curve must below the lower envelope curve of the protected equipment, that is, the two curves will never intersect. Therefore, the protective gap will act first to protect the insulation of the electrical equipment in any case.

6.1.2 Switching Impulse Breakdown Characteristics

The standard switching impulse voltage waveform specified in IEC standard is shown in Fig. 2.3a. Its front time $T_f = 250$ μs and the allowable deviation is ±20%. The time to half-value $T_2 = 2500$ μs and the allowable deviation is ±60%.

The air gap breakdown voltages under switching impulses are also of dispersion, and the U_{50} is also used to reflect the gap insulation strength. The switching impulse U_{50} is related to the electric field distribution and the voltage waveform. In uniform and slightly uneven electric field, the switching impulse U_{50} is the same with the power frequency breakdown voltage and the lightning impulse U_{50}, with small dispersion. In extremely nonuniform electric field, the switching impulse discharge characteristics of air gaps have polarity effect. Under the same gap distance, the negative breakdown voltage is higher than the positive one. Therefore, the positive discharge characteristics of air gaps should be mainly concerned for insulation coordination in practical engineering.

In extremely nonuniform electric field, the switching impulse discharge characteristics of air gaps are not only affected by the electric field distribution, but also related to the impulse voltage waveform. The double exponential wave is usually used to simulate the switching impulse voltage waveform, while the front time T_f and the time to half-value T_2 are used to characterize the waveform. For high voltage transmission systems, the front time of switching overvoltage is generally in the range of 50–1000 μs. But for EHV and UHV transmission systems, the front time can reach 1000–3000 μs [2]. A great deal of research work has been done on the influence of the switching impulse voltage waveform, especially the front time, on air gap discharge characteristics. With the variation of the front time T_f, the U_{50} has a minimum value. The relationship between U_{50} and T_f is an U-shape curve. Figure 6.4 shows the relationships between the positive switching impulse discharge voltages of rod-plane gaps and the front time of the impulse voltage [3, 4], which were obtained by discharge tests in the United States, Japan and Italy, etc. The minimum value of U_{50} is called "critical 50% breakdown voltage $U_{50,\text{crit}}$", the corresponding front time is called "critical wavefront time T_{cr}".

The U-shape curve is caused by the discharge delay as well as the formation and migration of space charges. When the wave front time is short, the voltage rises fast, and the discharge development requires a certain time delay, therefore the breakdown voltage is much higher than the static breakdown voltage. Hence, the breakdown voltage is higher than $U_{50,\text{crit}}$ when the front time is less than T_{cr}. When the wave front time is long, the voltage rises slowly, results in enough time for the formation and development of impulse corona and space charges in long air gap with extremely nonuniform electric field. The electric field distribution in the gap will be homogenized, thus increases the breakdown voltage.

As shown in Fig. 6.4, for a given gap structure, T_{cr} will gradually increase with the gap distance d. T_{cr} has an approximately linear relation with d. The relationships between T_{cr} and d for rod-plane gaps under positive and negative switching impulses, and for conductor-plane gaps under positive switching impulses are respectively given in [5–7], which can be expressed as

$$\begin{aligned} &\text{rod-plane gap (positive)}\ T_{\text{crit}} = 50(d-1) \approx 50d \\ &\text{rod-plane gap (negative)}\ T_{\text{crit}} = 10d \\ &\text{line-plane gap (positive)}\ T_{\text{crit}} = 35d \end{aligned} \tag{6.3}$$

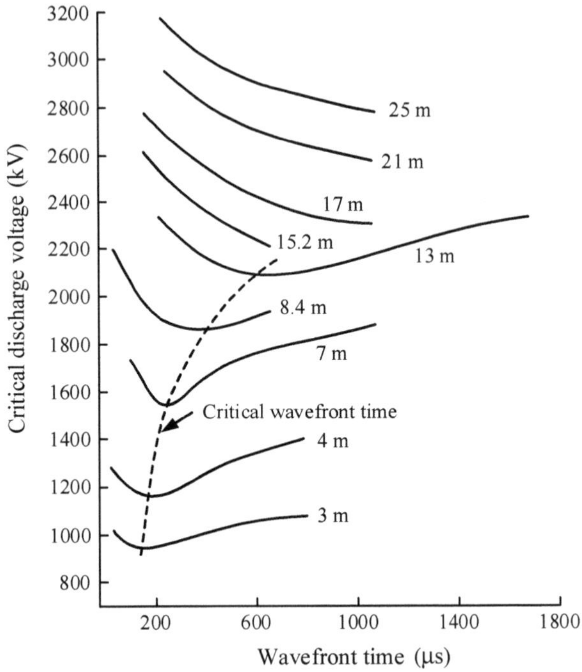

Fig. 6.4 Relationships between positive switching impulse discharge voltages of rod-plane gaps and the front time

Within the range of certain wave front time, the switching impulse breakdown voltage of the air gap is even lower than the power frequency breakdown voltage, which must be taken into account for air clearance determination of electrical equipment. In EHV and UHV transmission systems with the rated voltage greater than 220 kV, the insulation should be designed according to the electrical characteristics under switching overvoltage. By a large number of experimental studies, many scholars have put forward the empirical formulas between the switching impulse discharge voltage and the rod-plane gap distance.

In 1975, G. Gallet, et al. carried out the positive switching impulse discharge tests of rod-plane gaps with the length of 1–23 m. According to the experimental results, they proposed the relation between the $U_{50,\text{crit}}$ under positive switching impulses and the rod-plane gap distance, usually called EDF formula [8], which can be expressed as

$$U_{50,\text{crit}+} = 3400/(1 + 8/d) \qquad 1\,\text{m} \le d \le 23\,\text{m} \tag{6.4}$$

However, R. Cortina, et al. pointed out that [7] Eq. 6.4 is only applicable to the case of $d < 15$ m. When extrapolated to a larger gap, the calculated value of Eq. 6.4 will be smaller than the actual value. The relative error will be as high as 15% for $d = 30$ m. Based on the experimental data of longer air gaps, R. Cortina, et al. put forward the calculation formula of $U_{50,\text{crit+}}$ in the range of 13–30 m, which can be expressed as

$$U_{50,\text{crit+}} = 1400 + 55d \qquad 13\,\text{m} \leq d \leq 30\,\text{m} \tag{6.5}$$

In addition, R. Cortina et al. gave a formula to calculate the $U_{50,\text{crit}}$ of rod-plane gaps under the negative switching impulses when the gap distance d is in the range of 2–15 m:

$$U_{50,\text{crit}-} = 1180d^{0.45} \qquad 2\,\text{m} \leq d \leq 15\,\text{m} \tag{6.6}$$

In 1984, I. Kishizima, et al. in Central Research Institute of Electric Power Industry (CRIEPI) in Japan proposed a calculation formula for $U_{50,\text{crit+}}$ of rod-plane gaps, which is applicable to 1–25 m gap and can be expressed as [9]:

$$U_{50,\text{crit+}} = 1080\ln(0.46d + 1) \qquad 1\,\text{m} \leq d \leq 25\,\text{m} \tag{6.7}$$

Equation 6.7 has been recommended by IEC standard [10].

In 1989, based on the continuous leader model and some assumptions, F. Rizk deduced the relation between the $U_{50,\text{crit}}$ of rod-plane gaps and the gap distance d. For the gaps of $d \geq 4$ m, it can be expressed as [11, 12]:

$$U_{50} = \frac{1830 + 59d}{1 + 3.89/d} + 92 \qquad 4\,\text{m} \leq d \leq 25\,\text{m} \tag{6.8}$$

Figure 6.5 shows the relationship between the critical discharge voltages of rod-plane gaps and the gap distance derived from the EDF formula, the CRIEPI formula and the Rizk formula. It can be seen that the calculation results of the 3 empirical formulas are close for $d < 17$ m, but the differences increase gradually after $d > 17$ m. The saturation trend of the EDF formula is the most significant, and the discharge voltage calculated by Rizk formula has the largest increase gradient with the gap distance.

In 1976, G. Carrara, et al. proposed the concept of critical radius by analyzing the discharge characteristics of rod-plane gaps and conductor-plane gaps [13]. Taking rod-plane gaps for example, as shown in Fig. 6.6a, given a gap distance d, if the curvature radius of the rod tip is less than a critical value R_{cr}, the U_{50} of the rod-plane gaps will approximately remain unchanged under the positive switching impulse voltage with the critical wave front. For $R > R_{\text{cr}}$, U_{50} will increase obviously with the increase of R. The critical radius R_{cr} is related to the gap distance d. The formulas used to calculate R_{cr} of rod-plane and conductor-plane gaps under positive switching impulse voltage is given in [5]:

$$\begin{array}{ll} \text{Rod-plane gap} & R_{\text{cr}} = 0.38(1 - \text{e})^{-d/5} \\ \text{Conductor-plane gap} & R_{\text{cr}} = 0.037\ln(1 + d) \end{array} \tag{6.9}$$

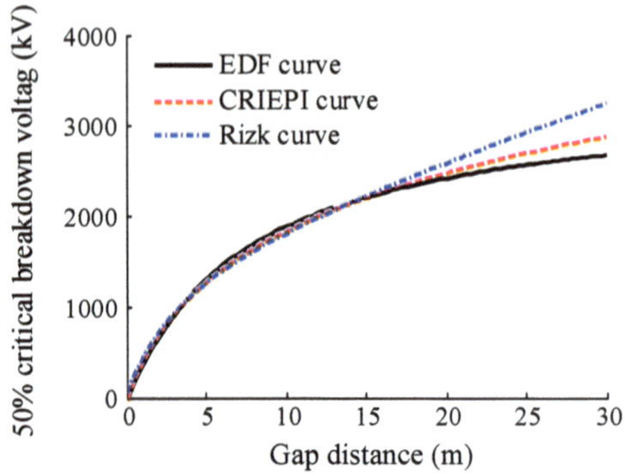

Fig. 6.5 Relation between the critical breakdown voltages of rod-plane gaps and the gap distance

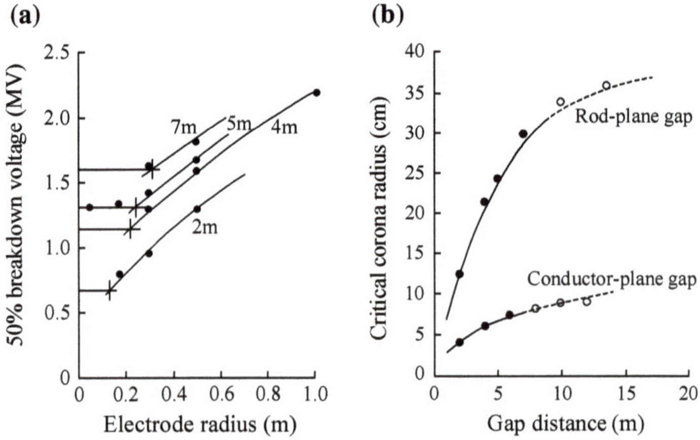

Fig. 6.6 Critical corona radius characteristics of rod (conductor)-plane gaps: **a** relation between U_{50} and R, **b** relation between R_{cr} and d

For positive discharge, R_{cr} will gradually become saturated with the increase of d, as shown in Fig. 6.6b. The saturation values of the critical radius for the rod electrode and the conductor are about 38 cm and 10 cm, respectively [14].

For the discharge characteristics of air gaps used in transmission and transformation projects, L. Paris proposed the concept of "gap factor" in 1966 [15, 16]. It is considered that the U_{50} of an engineering gap is equal to the product of the positive switching impulse discharge voltage of the rod-plane gap and the gap factor k. Based on gap factor method, L. Paris conducted the discharge characteristic tests of rod-rod, rod-plane, conductor-plane, conductor-rod, conductor-tower window, and conductor-structure gaps, etc. under the positive 120/4000 μs switching impulse voltages. He proposed the U_{50} calculation formula for different gap configurations under the positive switching impulse voltage, which can be expressed as:

$$U_{50} = k \cdot 500d^{0.6} \tag{6.10}$$

The gap factors k of different configurations are also given. For rod-plane gaps, $k = 1$. For conductor to grounded rod gaps, $k = 1.9$. For conductor-plane, conductor-tower window, rod-rod gaps, etc., the values of k are in the range of 1.05–1.9.

In 1992, CIGRE SC 33-07 working group published the guidelines for the evaluation of the dielectric strength of external insulation [17], and proposed a method to calculate the gap factor considering the effects of gap configuration, gap length, and electrode shape, etc. This brochure also gave the calculation formulas and typical values of the gap factors for some actual configurations, such as the conductor-cross arm, conductor-window and rod-rod structure, etc.

6.2 Switching Impulse Discharge Voltage Prediction

In order to verify the feasibility of the air insulation prediction model in switching impulse discharge voltages of long air gaps, the electric field features and the impulse voltage waveform features are both taken as the input parameters of the SVM model, and it is used to predict the discharge voltages of long air gaps with different electrode structures, submitted to different switching impulses.

6.2.1 Switching Impulse Discharge Voltage Prediction of Rod-Plane and Rod-Rod Gaps

6.2.1.1 Sample Data

The positive and negative switching impulse discharge tests of rod-plane and rod-rod long air gaps were carried out in [18]. The experimental data are cited as samples to verify the effectiveness of the proposed prediction method.

The rod electrode is made of mild steel with the tip 12.5 mm by 12.5 mm in area. The plane electrode is a galvanized steel sheet with 13 m by 13 m in area. The impulse voltage waveform is (80–550)/(1800–2200) μs. All tests were carried out outdoors. The high voltage electrode was set up more than 15 m from the surrounding steel structures. Switching impulse voltage was applied more than 50 times for each gap to obtain the U_{50}. The relative air density during the tests ranges from 0.92 to 0.98, and the experimental data were corrected to the atmospheric conditions at 101.3 kPa and 20 °C, but not corrected for humidity conditions. The experimental results are shown in Tables 6.1 and 6.2, where d is the gap distance, T_f is the front time of the impulse voltage waveform, and σ is the standard deviation of the experimental data. 2 data are selected as training samples from Tables 6.1 and 6.2, while all the data are taken as test samples. The positive and negative switching impulse U_{50} of rod-plane and rod-rod gaps are predicted respectively.

Table 6.1 Switching impulse U_{50} of rod-plane gaps (sample data)

Sample classification	Positive polarity				Negative polarity			
	d (m)	T_f (μs)	U_{50} (kV)	σ (%)	d (m)	T_f (μs)	U_{50} (kV)	σ (%)
Training samples	9.0	80	2200	3.6	3.0	80	2040	6.3
	9.0	110	2070	4.0	3.0	180	2300	6.3
Test samples	5.0	180	1340	6.7	1.5	180	1500	4.0
	7.0	180	1650	6.7	2.0	180	1800	5.5
	8.0	180	1790	6.1	2.5	180	2080	3.4
	9.0	180	1950	4.9	3.0	110	2060	3.4
	9.0	250	1930	4.9	3.0	250	2340	6.0
	9.0	550	1900	4.8	3.0	550	2340	5.9
	10.0	180	2050	4.9	3.5	180	2480	6.2
	11.0	180	2100	4.8	4.0	180	2620	5.6
	12.0	180	2180	6.6	–	–	–	–
	13.0	180	2300	6.8	–	–	–	–

Table 6.2 Switching impulse U_{50} of rod-rod gaps (sample data)

Sample classification	Positive polarity				Negative polarity			
	d (m)	T_f (μs)	U_{50} (kV)	σ (%)	d (m)	T_f (μs)	U_{50} (kV)	σ (%)
Training samples	5.5	80	2150	5.6	3.0	80	1940	5.2
	5.5	110	2100	4.8	3.0	110	1960	5.3
Test samples	4.0	180	1690	5.3	1.5	180	1110	5.5
	5.0	180	1940	3.6	2.0	180	1450	5.4
	5.5	180	2080	6.7	2.5	180	1790	4.5
	5.5	250	2050	5.5	3.0	180	2020	4.5
	5.5	550	2050	5.5	3.0	250	2030	4.5
	6.0	180	2180	6.8	3.0	550	2040	5.3
	7.0	180	2490	4.9	3.5	180	2330	3.2
	7.5	180	2490	4.0	4.0	180	2620	4.0
	–	–	–	–	4.5	180	2780	3.8
	–	–	–	–	5.0	180	2970	3.6

6.2.1.2 Prediction Results and Analysis

The 20-d electric field features and 10-d impulse voltage waveform features are taken together as input parameters of the SVM model. For the electric field features, the 8 features related to 24 and 7 kV/cm are rejected from the total 28-d. The 10-d impulse voltage waveform features include the 5 basic features and the 5 additional features related to 85% U_{max}. Under the sense of 3-CV, GA is used to optimize the SVM parameters. The population quantity is set as 20, the maximum generation is 200, the crossover probability is 0.9, and the search ranges of the penalty factor C and the kernel parameter γ are respectively [10, 500] and [0.005, 0.25]. The optimal parameters and error indexes of the SVM prediction results are shown in Table 6.3. The e_{MAPE}s of the 4 test sample set are respectively 3.6%, 3.25%, 3.5% and 3.8%.

The U_{50} prediction results of rod-plane and rod-rod gaps under positive and negative switching impulse voltages are compared with the experimental data, as shown in Figs. 6.7, 6.8, 6.9 and 6.10, while the predicted results of training samples are also plotted in these figures.

It can be seen from Figs. 6.7, 6.8, 6.9 and 6.10 that the predicted values agree well with the experimental data, and the predicted results of most samples are within the standard deviation range of the experimental data. The results indicate that the SVM model has higher prediction accuracy.

Table 6.3 Optimal parameters and error indexes of the SVM prediction model

Parameters	Rod-plane gap		Rod-rod gap	
	Positive polarity	Negative polarity	Positive polarity	Negative polarity
C	13.4294	16.8959	332.3200	23.6726
γ	0.0915	0.0965	0.2130	0.1479
e_{MSE}	24.4052	26.8715	35.8847	30.1377
e_{MAPE}	0.0360	0.0325	0.0350	0.0380
e_{MSPE}	0.0149	0.0131	0.0158	0.0151

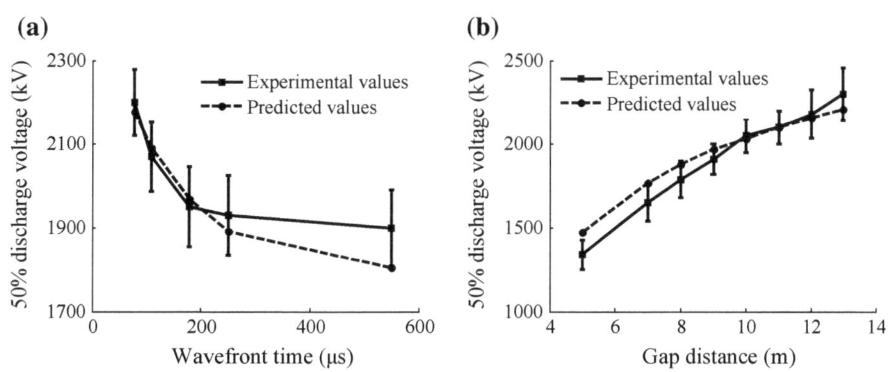

Fig. 6.7 Comparison between the predicted and experimental U_{50} of rod-plane gaps under positive switching impulse voltages: **a** $T_f = 180$ μs, $d = 5$–13 m, **b** $d = 9$ m, $T_f = 80$–550 μs

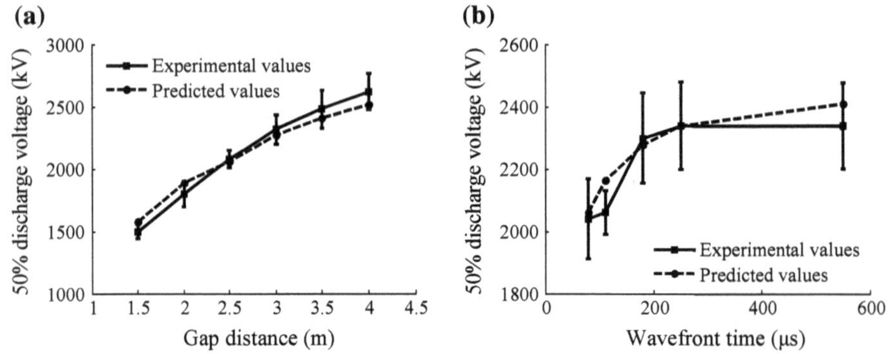

Fig. 6.8 Comparison between the predicted and experimental U_{50} of rod-plane gaps under negative switching impulse voltages: **a** $T_{\rm f} = 180$ μs, $d = 1.5$–4 m, **b** $d = 3$ m, $T_{\rm f} = 80$–550 μs

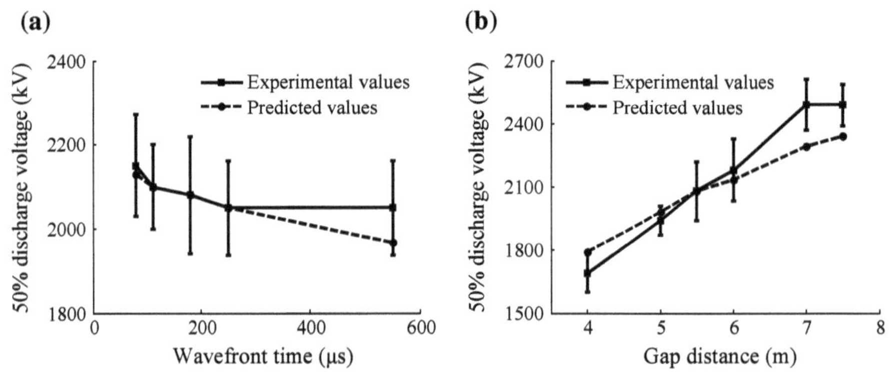

Fig. 6.9 Comparison between the predicted and experimental U_{50} of rod-rod gaps under positive switching impulse voltages: **a** $T_{\rm f} = 180$ μs, $d = 4$–7.5 m, **b** $d = 5.5$ m, $T_{\rm f} = 80$–550 μs

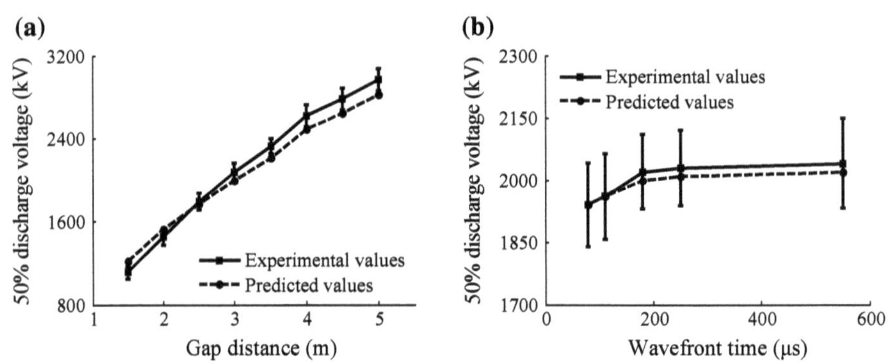

Fig. 6.10 Comparison between the predicted and experimental U_{50} of rod-rod gaps under negative switching impulse voltages: **a** $T_{\rm f} = 180$ μs, $d = 1.5$–5 m, **b** $d = 3$ m, $T_{\rm f} = 80$–550 μs

In addition, it can be seen from Fig. 6.8b that the experimental U_{50} under the negative switching impulse voltage with $T_f = 250$ μs is equal to that of $T_f = 550$ μs. As is shown in Fig. 6.9, when $T_f = 180$ μs, the experimental U_{50} of $d = 7$ m is equal to that of $d = 7.5$ m, and when $d = 5.5$ m, the experimental U_{50} of $T_f = 250$ μs is equal to that of $T_f = 550$ μs. This may be due to the lack of humidity correction for the test data in [18]. However, the SVM prediction results corresponding to the above samples show obvious differences, which indicates that the predicted values are more consistent with the actual situation than the experimental data [19].

6.2.2 *Hybrid Prediction of Switching Impulse Discharge Voltages of Different Gap Structures*

6.2.2.1 Sample Data

Discharge tests of rod-plane, rod-rod and rod-conductor long air gaps were conducted in [20] subjected to negative switching impulse voltages of 20/2500 and 80/2500 μs. The experimental results were taken as training and test samples of the predictive model. The schematic diagrams of the three gap arrangements are shown in Fig. 6.11. The high voltage rod electrode is a 15 m long circular iron rod whose diameter is 6 cm, with a copper ball on the tip, 8 cm in diameter. The plane electrode is a 20 m × 20 m iron plate placed on 50 m × 50 m grounded iron net. The grounded rod electrode is a 4.5 m long circular iron rod whose diameter is 6 cm. The conductor is a scale model of 8-bundle conductor, the diameter of each sub-conductor is 2.7 mm, the distance between sub-conductors is 32 mm, and the distance above the earth is $h = 4.5$ m. The gap distance d is 1–10 m, and the test results were corrected to standard atmospheric conditions.

The experimental results are shown in Table 6.4, where T_f is the front time, d is the gap length and σ is the standard deviation. 6 test data of rod-plane and rod-rod gaps with three gap lengths and two voltage waveforms were selected as training samples. The predictive model trained by the 6 training samples is used to predict the U_{50} of the other gaps with different gap geometries and under different impulse voltages, altogether 33 test samples.

Table 6.4 Experimental U_{50} of rod-plane, rod-rod and rod-conductor gaps under switching impulse (sample data): (**a**) training sample set, (**b**) experimental data of test sample set

Gap structure	T_f (μs)	d (m)	U_{50} (kV)	σ (%)
a				
Rod-plane gap	20	1	950	9.5
		5	2393	4.4
		10	3672	7.5

(continued)

Table 6.4 (continued)

Gap structure	T_f (μs)	d (m)	U_{50} (kV)	σ (%)
Rod-rod gap	80	1	756	2.4
		5	2557	4.2
		10	3607	4.7
b				
Rod-plane gap	20	1	950	9.5
		2	1524	4.2
		3	1901	4.1
		4	2227	2.9
		5	2393	4.4
		6	2826	3.4
		8	3246	4.9
		10	3672	7.5
Rod-plane gap	80	1	938	5.7
		2	1505	5.2
		3	1915	4.4
		4	2146	4.2
		5	2404	5.5
		6	2616	2.9
		8	2941	2.6
		10	3313	2.6
Rod-rod gap	20	1	697	4.9
		2	1335	3.3
		3	1881	4.4
		4	2322	2.6
		5	2552	3.3
		6	2898	3.2
		8	3458	3.2
		10	3979	3.3
Rod-rod gap	80	1	756	2.4
		2	1281	2.9
		3	1847	5.1
		4	2247	4.1
		5	2557	4.2
		6	2707	2.2
		8	3182	2.9
		10	3607	4.7

(continued)

Table 6.4 (continued)

Gap structure	T_f (μs)	d (m)	U_{50} (kV)	σ (%)
Rod-conductor gap	80	1	759	3
		2	1398	4
		3	1885	5.5
		5	2435	3.9
		6	2737	3.5
		8	3150	2.5
		10	3578	2.6

6.2.2.2 Electric Field Calculation and Feature Extraction

In order to extract the electric field features of every long air gap in Table 6.4, their electric field distributions were calculated by a finite element analysis software. The high voltage rod electrode was applied 1 V and the grounded electrode was applied zero potential. Taking the 5 m long gaps for example, the electric field distributions of rod-plane, rod-rod and rod-conductor air gaps along the shortest discharge path are shown in Fig. 6.12a. It can be seen that the field strengths on the shortest path are very

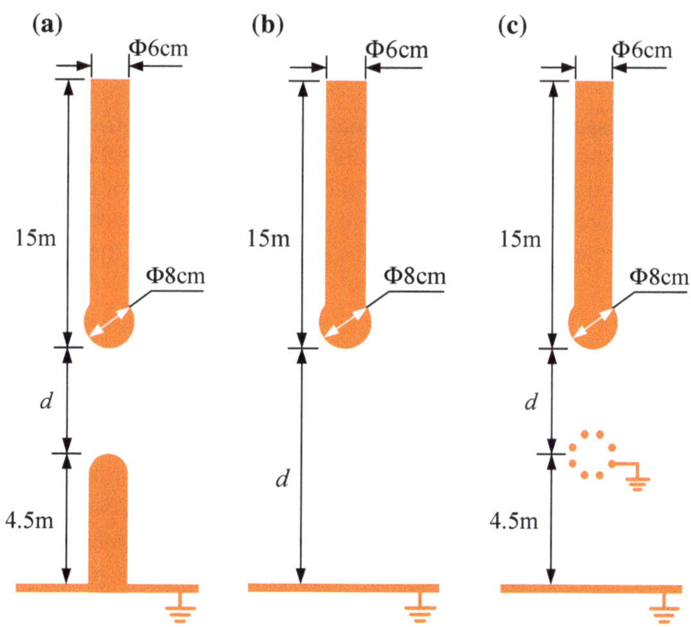

Fig. 6.11 Structure diagrams of rod-plane, rod-rod and rod-conductor air gaps: **a** rod-plane gap, **b** rod-rod gap, **c** rod-conductor gap

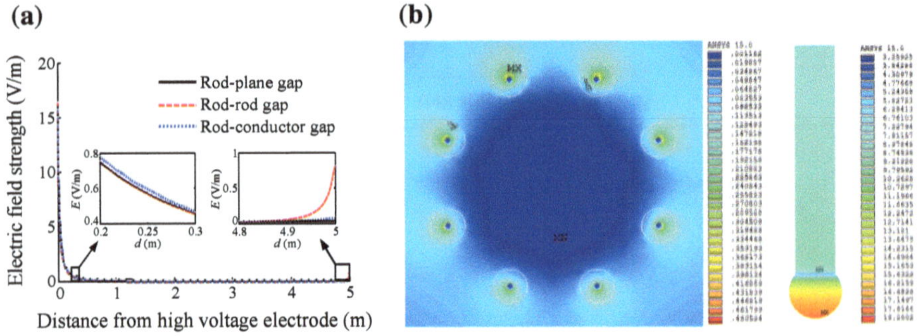

Fig. 6.12 Electric field distributions of the three kinds of long air gaps ($d = 5$ m): **a** electric field distributions along the shortest discharge path, **b** electric field distribution on the electrode surface of rod-conductor gap

close for the three gap types. Around the high voltage electrode, the field strength of rod-conductor gap is a bit larger than that of rod-plane and rod-rod gap, while the latter two are basically identical. Near the grounded electrode, the electric field strength of rod-rod gap is the largest, rod-conductor gap second and the rod-plane gap approaches zero.

Taking the rod-conductor gap as example, the cloud chart of the electric field distribution on the electrode surface is shown in Fig. 6.12b. The maximum field strength appears on the surface of the high voltage rod electrode, and there is also field concentration around the 8-bundle conductor, but the field strength is much lower.

By post-processing of the FEM calculation results, the electric field features are extracted. In addition, the 20/2500 and 80/2500 μs impulse voltage waveform features are calculated according to their definitions and formulas of these features. The electric field features and impulse voltage waveform features constitute the energy storage feature set and they are taken as the input parameters of the SVM model after normalization.

6.2.2.3 Prediction Results and Analysis

Based on 3-CV, the improved GS algorithm is used to optimize the parameters of the SVM model, and the optimal parameters are $C = 119.4282$, $\gamma = 0.25$, the highest classification accuracy of the training samples is 96.8254%. In order to get the U_{50} prediction value of each test sample quickly and improve the computational efficiency of the prediction model, the golden section search method is used for iterative calculation and prediction [21].

For an air gap belong to the test samples, set the estimated flashover voltage in the range $[U_{\min}, U_{\max}]$, the golden section method is used to search the predicted result U_{p}. Set the first applied voltage U_{t} as $U_{\text{t1}} = U_{\max} - 0.618 \times (U_{\max} - U_{\min})$, the electric field features and impulse voltage waveform features are extracted using the same method as training samples. After feature normalization, these energy storage features are input to the optimal prediction model. If the predictive model outputs 1, it means U_{t1} is larger or equal to the critical flashover voltage and therefore the search range of U_p is narrowed to $[U_{min}, U_{t1}]$. If the model outputs -1, it means that U_{t1} is not enough to make the gap breakdown, then, another voltage $U_{t2} = U_{min} + 0.618 \times (U_{max} - U_{min})$ is applied on the gap for prediction. If the model outputs 1, then the search range is adjusted to $[U_{t1}, U_{t2}]$. On the contrary, if the model under U_{t2} outputs -1, the search scope of U_p should be narrowed to $[U_{t2}, U_{max}]$. Through the above-mentioned approach, after multiple iterations, U_{max} and U_{min} are updated continuously and the search range of U_p is narrowed until $U_{max} - U_{min} < \varepsilon$, where ε is the convergence precision. Then the predicted result of the critical flashover voltage is $U_p = (U_{min} + U_{max})/2$. The implementation process of the above methods can be summarized in Fig. 6.13.

The 50% discharge voltages of test samples are predicted one by one. The initial range of the applied voltage is set as $U_{min} = 500$ kV, $U_{max} = 4000$ kV, and the convergence precision $\varepsilon = 1$ V. The predicted value of U_{50} is searched by the golden section method. After multiple iterations, the results are shown in Table 6.5, where δ is the relative error between the predicted value U_p and the test value U_t.

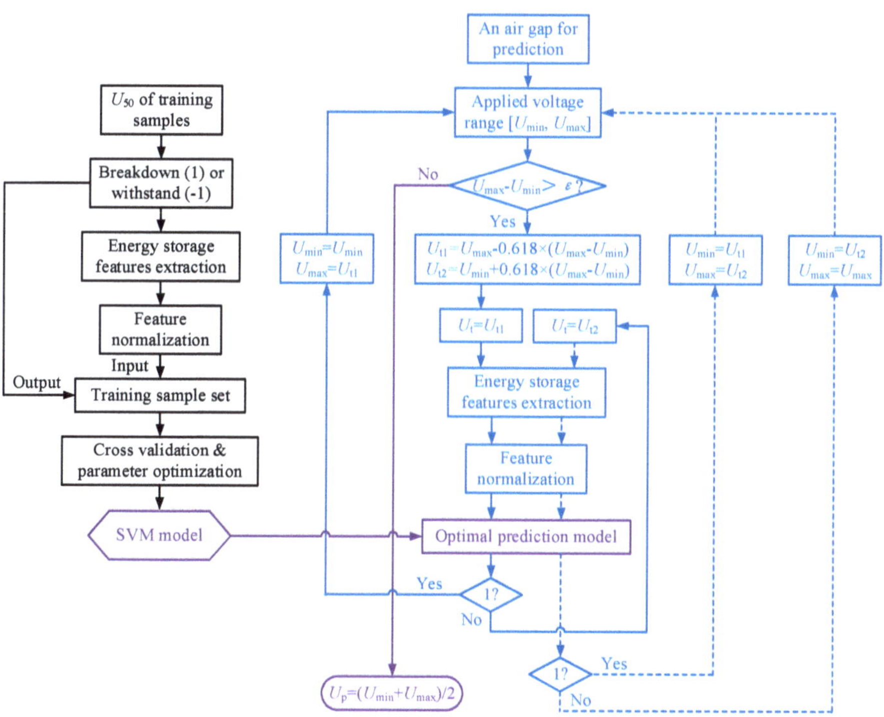

Fig. 6.13 Insulation prediction flow chart based on the golden section search method

Table 6.5 Switching impulse U_{50} prediction results of rod-plane, rod-rod and rod-conductor gaps: (**a**) rod-plane gap, (**b**) rod-rod gap, (**c**) rod-conductor gap

T_f (μs)	d (m)	U_t (kV)	U_p (kV)	δ (%)
a				
20	2	1524	1381	−9.4
	3	1901	1724	−9.3
	4	2227	2063	−7.4
	6	2826	2711	−4.1
	8	3246	3246	0
80	1	938	849	−9.5
	2	1505	1396	−7.2
	3	1915	1819	−5.0
	4	2146	2176	1.4
	5	2404	2484	3.3
	6	2616	2756	5.4
	8	2941	3187	8.4
	10	3313	3526	6.4
b				
80	2	1281	1385	8.1
	3	1847	1874	1.5
	4	2247	2245	−0.1
	6	2707	2823	4.3
	8	3182	3254	2.3
20	1	697	825	18.4
	2	1335	1405	5.2
	3	1881	1800	−4.3
	4	2322	2167	−6.7
	5	2552	2506	−1.8
	6	2898	2815	−2.9
	8	3458	3322	−3.9
	10	3979	3688	−7.3
c				
80	1	759	768	1.2
	2	1398	1379	−1.4
	3	1885	1812	−3.9
	5	2435	2455	0.8
	6	2737	2710	−1.0
	8	3150	3135	−0.5
	10	3578	3464	−3.2

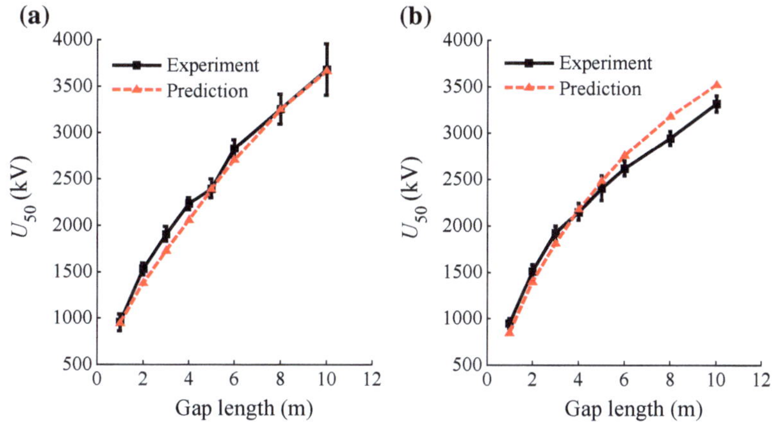

Fig. 6.14 Comparison between the predicted and experimental U_{50} of rod-plane gaps: **a** T_f = 20 μs, **b** T_f = 80 μs

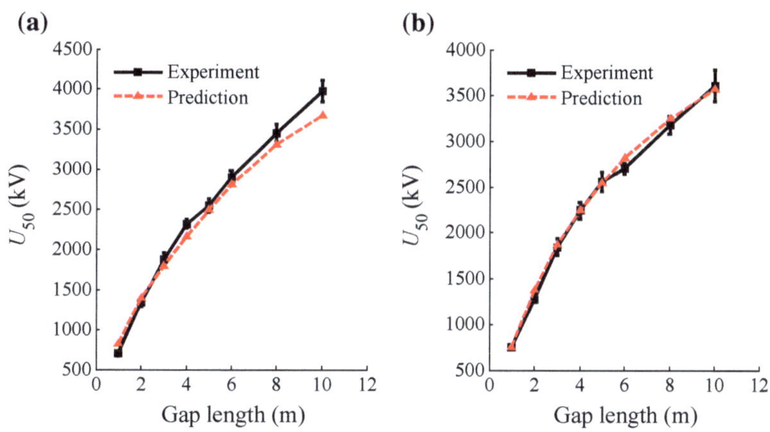

Fig. 6.15 Comparison between the predicted and experimental U_{50} of rod-rod gaps: **a** T_f = 20 μs, **b** T_f = 80 μs

The predicted and experimental results of rod-plane, rod-rod and rod-conductor air gaps are compared in Fig. 6.14, Fig. 6.15 and Fig. 6.16, respectively. For better comparisons, the predicted results of the training samples were also plotted in Figs. 6.14a and 6.15b. It can be seen that most of the predicted values are in good accordance with the experimental data, while the relative errors are within acceptable range for engineering applications.

The error indexes of the three gap types under two voltage waveshapes, including e_{MSE}, e_{MAPE} and e_{MSPE} are listed in Table 6.6. The e_{MAPE}s of rod-plane gaps under switching impulse voltages of 20/2500 μs and 80/2500 μs are respectively 6.0% and 5.8%, while 6.3% and 3.2% for rod-rod gaps. For rod-conductor gaps under

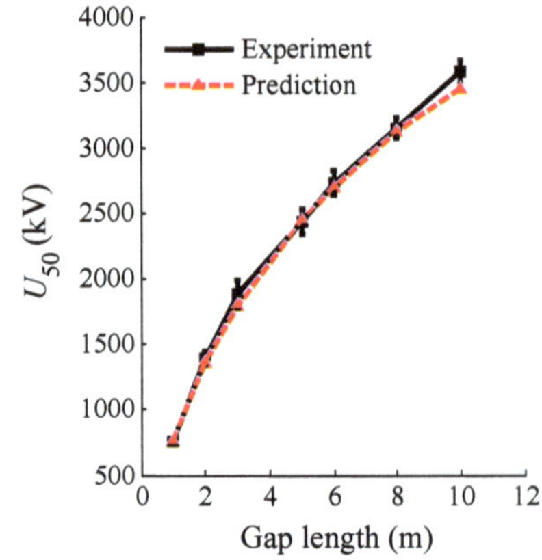

Fig. 6.16 Comparison between the predicted and experimental U_{50} of rod-conductor gaps

Table 6.6 Error indexes of U_{50} prediction results of rod-plane, rod-rod and rod-conductor gaps

Error indexes	Rod-plane gap		Rod-rod gap		Rod-conductor gap
	$T_f = 20\ \mu s$	$T_f = 80\ \mu s$	$T_f = 20\ \mu s$	$T_f = 80\ \mu s$	$T_f = 80\ \mu s$
e_{MSE}	60.63	50.28	50.63	34.75	20.26
e_{MAPE}	0.060	0.058	0.063	0.032	0.017
e_{MSPE}	0.031	0.022	0.028	0.019	0.008

negative 80/2500 μs impulse voltage, the predicted results coincide very well with the experimental values, and the e_{MAPE} of the predicted results is only 1.7%, the maximum absolute percentage error is 3.9%.

The results indicate that the proposed model is effective for U_{50} prediction of long air gaps. Trained by a few known experimental values, the SVM model can be used to predict the critical flashover voltages of long air gaps with different configurations and voltage waveforms. The method can be used to replace the costly and time-consuming long air gap discharge tests, so as to greatly reduce the test work and cost.

6.3 Lightning Impulse Discharge Voltage Prediction

The American Institute of Electrical Engineers (AIEE) carried out the lightning impulse discharge tests of rod-rod air gaps in 1930s, under the waveforms of 1/5 and 1.5/40 μs [22]. The experimental data given in [22] were obtained under the atmospheric environment of $t = 25\ °C$, $p = 101.3$ kPa and $h = 15\ g/m^3$. IEEE Std 4-1995 [23] recommended the arrangement of a rod-rod gap for voltage measurement, as shown in Fig. 6.17, and gave the critical sparkover voltages of these rod-rod gaps with different gap lengths, subjected to the lightning impulses of 1.2/5 and 1.2/50 μs. These voltage values are under standard atmospheric conditions. The rod gap con-

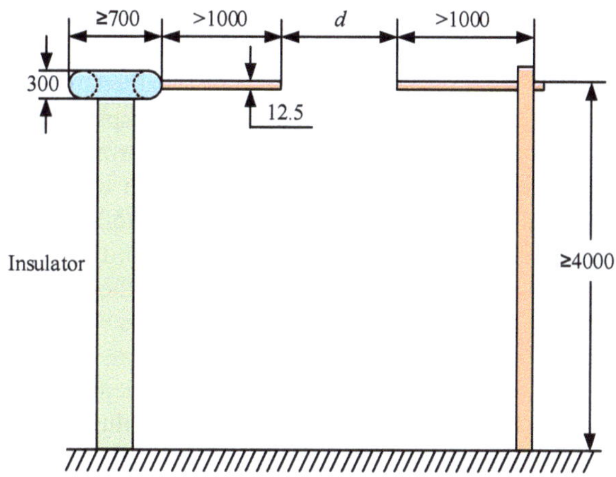

Fig. 6.17 Arrangement of the standard rod-rod gap (all dimensions are in mm)

sists of two 12.5 mm square rod electrodes arranged horizontally. The experimental results under the above-mentioned lightning impulses, with 4 kinds of voltage waveforms, are taken as sample data to validate the validity of the proposed method for lightning impulse discharge voltage prediction.

6.3.1 Positive Lightning Impulse Discharge Voltage Prediction

6.3.1.1 Experimental Data and Sample Selection

The rod-rod gaps with the gap distance ranging from 20 to 152.4 cm are selected as samples from the AIEE experimental results and the IEEE Std4-1995. In order to make the predictive model generalize to various gap configurations and voltage waveforms, the selected training sample set contains the experimental data of rod-rod gaps with different gap lengths and voltage waveforms, as shown in Table 6.7, altogether 7 samples. The values with * are caused by unstable conditions [23], and the average value is taken as the corresponding experimental discharge voltage.

Table 6.7 Training sample set of positive lightning impulse discharge voltage

Waveform	*d* (cm)	*U* (kV)	Waveform (μs)	*d* (cm)	*U* (kV)
1.2 × 50 μs	20	154–161*	1.2 × 5	30	277
	50	339	1.5 × 40	76.2	505
	100	625	1 × 5	127	1035
	140	850	–	–	–

Other gaps with different gap distances and different impulse voltage waveforms are taken as test samples for flashover voltage prediction and the predicted results are compared with the experimental data obtained by IEEE Std 4-1995 and AIEE. There are altogether 38 test samples of rod-rod gaps, as shown in Table 6.8. Furthermore, in order to validate the generalization performance of the prediction model for other gap types, 7 samples of sphere-plane gap under 1/50 μs lightning impulse [24] and 4 samples of rod-plane gap under 1.2/50 μs waveshape [25] are also taken as the test samples. For the sphere-plane gap, the sphere diameter is 2.5 cm and the plane electrode is a 60 cm diameter disc, the gap length ranges from 10 to 50 cm. The experimental data of sphere-plane gaps were obtained under normal laboratory atmospheric conditions and no correction was made [24]. The rod-plane gap consists of a cylindrical brass rod with a hemispherical tip and a grounded aluminium plane, the rod diameter is 2.2 cm, and the size of the plane electrode is 100 cm × 200 cm.

Table 6.8 Experimental data of test samples for positive lightning impulse discharge voltage prediction

Rod-rod (1.2/50 μs)		Rod-rod (1.2/5 μs)		Rod-rod (1.5/40 μs)	
d (cm)	*U* (kV)	*d* (cm)	*U* (kV)	*d* (cm)	*U* (kV)
25	184	20	188	20.32	162.5
30	217	25	234	22.86	176.5
35	250	35	320	25.4	190
40	281	40	362	38.1	275
45	309	45	405	50.8	350
60	392	50	445	101.6	650
70	450	60	525	127	800
80	510	70	605	152.4	945
90	570	80	690	–	–
120	735	90	765	–	–
–	–	100	845	–	–
–	–	120	990	–	–
–	–	140	1150	–	–
Sphere-plane (1/50 μs)		Rod-plane (1.2/50 μs)		Rod-rod (1/5 μs)	
d (cm)	*U* (kV)	*d* (cm)	*U* (kV)	*d* (cm)	*U* (kV)
10	65	12.5	95	20.32	187
15	90	25	163	25.4	233
20	113	37.5	223	38.1	340
25	140	50	278	50.8	440
30	170	–	–	76.2	640
40	235	–	–	101.6	835
50	267	–	–	152.4	1230

The data of rod-plane gaps were corrected to standard atmospheric conditions [25]. The experimental data of the above-mentioned test samples are shown in Table 6.8, where d is the gap length and U is the discharge voltage.

6.3.1.2 Electric Field Calculation and Feature Extraction

The finite element models of rod-rod, sphere-plane and rod-plane gaps are established for electric field calculation and feature extraction. Taking the 30 cm rod-rod gap for example, a three-dimensional finite element model is established according to the electrode sizes shown in Fig. 6.17. The high voltage rod electrode is applied unit voltage 1 V and the grounded rod is applied 0 V, its electric field distribution is shown in Fig. 6.18. The discharge channel is a square region between two electrodes, the length of side is twice that of the rod electrode. The shortest discharge path is the connecting line from the cross section center of the high voltage rod to that of the grounded rod [26]. For the hemispherical rod-plane and sphere-plane configurations, the discharge channel is the cylindrical region between the rod or sphere electrode and the plane electrode, and the radius of the cylinder is the same with the rod or sphere electrode.

By post-processing of the FEM calculation results, the electric field features are extracted and calculated. In addition, the double exponential functions of lightning

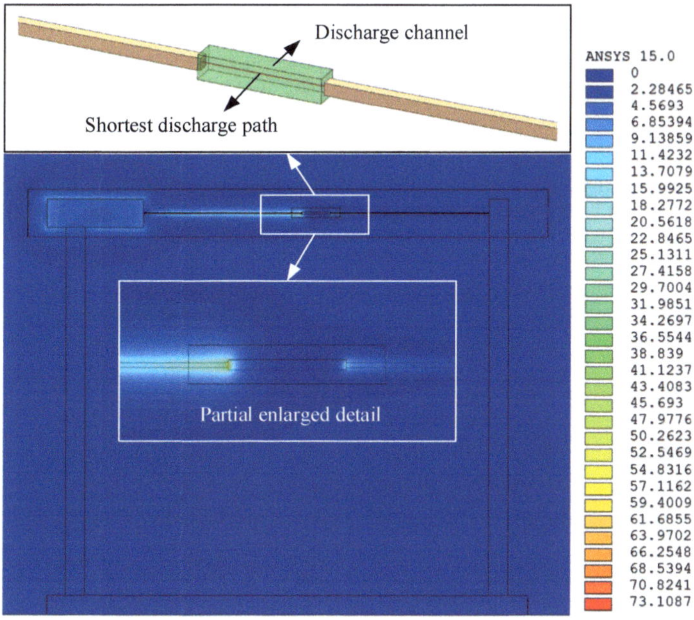

Fig. 6.18 Electric field distribution of the standard rod-rod gap (d = 30 cm) and the schematic diagram of the discharge channel and the shortest discharge path

impulses with different waveshapes can be solved by their constraint equations. Therefore, the impulse voltage waveform features of each sample can be obtained according to their calculation formulas. The electric field features and the impulse voltage waveform features are summarized to constitute the energy storage feature set after normalization, which are taken as the input parameters of the SVM model.

6.3.1.3 Prediction Results

The training samples are used to train the SVM model, and the improved GS algorithm is used for parameter optimization. The optimal parameters are $C = 19.6983$, $\gamma = 0.25$. The discharge voltage of each test sample is predicted one by one. The initial range of the applied voltage is set as $U_{\min} = 0$ kV and $U_{\max} = 1500$ kV. The convergence precision ε is set as 1 kV. The predicted results could be obtained after multiple iterative computations by golden section search method.

The positive lightning impulse discharge voltage prediction results of rod-rod, sphere-plane and rod-plane gaps are shown in Table 6.9, where U_{p} is the predicted discharge voltage, and σ is the relative error between the predicted value and the experimental value. Figure 6.19 summarizes the comparison of the discharge voltage prediction results of rod-rod gaps with the experimental values obtained by IEEE Std 4-1995 and AIEE. For better comparisons, the predicted results of the training samples are also plotted in Fig. 6.19.

It can be seen that the predicted values of the discharge voltage correlate well with the experimental data. The predicted results of all the test samples, under 4 different lightning impulse waveforms, have high accuracy. The relative errors of the predicted values are within 8% for most of the test samples, which are acceptable

Table 6.9 Positive lightning impulse discharge voltage prediction results of test samples: (**a**) Rod-rod gaps (1.2/50 and 1.2/5 μs), (**b**) rod-rod gaps (1.5/40 and 1/5 μs), (**c**) rod-plane and sphere-plane gaps

a

Rod-rod gaps (1.2/50 μs)				Rod-rod gaps (1.2/50 μs)			
d (cm)	U (kV)	U_{p} (kV)	σ (%)	d (cm)	U (kV)	U_{p} (kV)	σ (%)
25	184	183.7	−0.2	20	188	186.4	−0.9
30	217	217.9	0.4	25	234	218.4	−6.7
35	250	249.9	−0.04	35	320	305.6	−4.5
40	281	275.8	−1.9	40	362	342.0	−5.5
45	309	308.3	−0.2	45	405	384.8	−5.0
60	392	401.0	2.3	50	445	399.4	−10.3
70	450	467.7	3.9	60	525	504.5	−3.9
80	510	514.2	0.8	70	605	593.1	−2.0

(continued)

Table 6.9 (continued)

a

Rod-rod gaps (1.2/50 μs)				Rod-rod gaps (1.2/50 μs)			
d (cm)	*U* (kV)	U_p (kV)	*σ* (%)	*d* (cm)	*U* (kV)	U_p (kV)	*σ* (%)
90	570	576.5	1.1	80	690	657.8	−4.7
120	735	740.1	0.7	90	765	739.6	−3.3
–	–	–	–	100	845	805.7	−4.7
				120	990	959.0	−3.1
				140	1150	1092.8	−5.0

b

Rod-rod gaps (1.5/40 μs)				Rod-rod gaps (1/5 μs)			
d (cm)	*U* (kV)	U_p (kV)	*σ* (%)	*d* (cm)	*U* (kV)	U_p (kV)	*σ* (%)
20.32	162.5	162.1	−0.3	20.32	805.7	190.3	1.8
22.86	176.5	178.9	1.4	25.4	233	221.4	−5.0
25.4	190	189.1	−0.5	38.1	340	338.6	−0.4
38.1	275	278.5	1.3	50.8	440	437.0	−0.7
50.8	350	351.7	0.5	76.2	640	594.4	−7.1
101.6	650	661.6	1.8	101.6	835	850.6	1.9
127	800	797.0	−0.4	152.4	1230	1224.2	−0.5
152.4	945	954.6	1.0	–	–	–	–

c

Rod-plane gaps (1.2/50 μs)				Sphere-plane gaps (1/50 μs)			
d (cm)	*U* (kV)	U_p (kV)	*σ* (%)	*d* (cm)	*U* (kV)	U_p (kV)	*σ* (%)
12.5	95	81.7	−13.7	10	65	63.9	−1.7
25	163	157.7	−3.0	15	90	89.6	−0.4
37.5	223	227.2	2.0	20	113	117.1	3.6
50	278	290.2	4.3	25	140	144.9	3.5
–	–	–	–	30	170	172.1	1.2
				40	235	223.3	−5.0
				50	267	270.5	1.3

for engineering applications. It should be noted that the experimental data also have certain errors, which may be as high as ±8% [23]. In general, the errors of the predicted results are within acceptable range.

The comparison of the discharge voltage prediction results of rod-plane and sphere-plane gaps with the experimental data is shown in Fig. 6.20. It can be seen that the predicted values of sphere-plane gaps under 1/50 μs impulse agree well with the test results, and the relative errors are within 5%. The predicted results of rod-plane gaps under 1.2/50 μs impulse diverge a little from the experimental data in trend, but the relative errors are also within acceptable range. Except for the sample

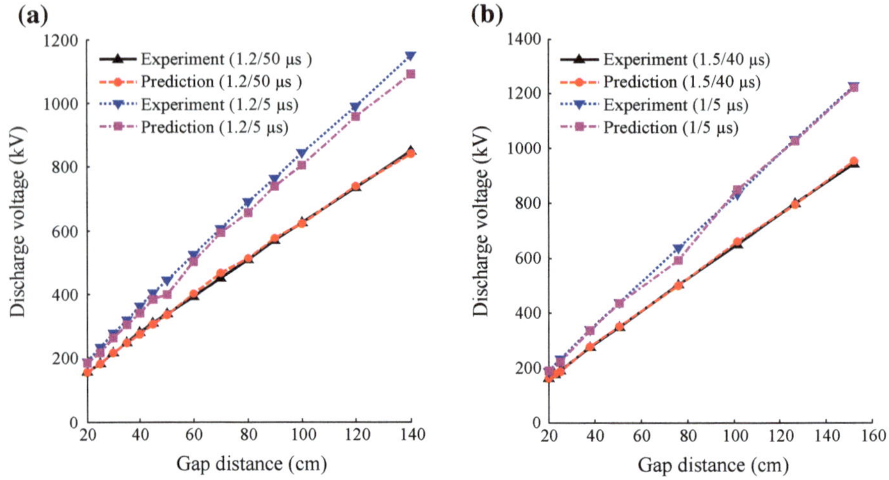

Fig. 6.19 Comparison of positive lightning impulse discharge voltage prediction results of rod-rod gaps with the experimental data: **a** 1.2/50 and 1.2/5 μs, **b** 1.5/40 and 1/5 μs

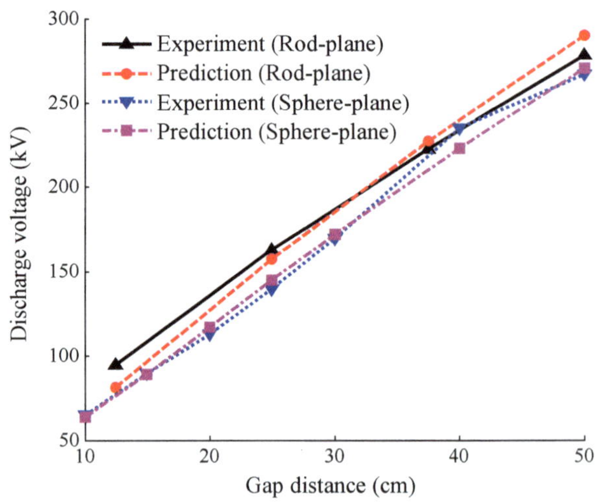

Fig. 6.20 Comparison of discharge voltage prediction results of rod-plane and sphere-plane gaps with the experimental data

of $d = 12.5$ cm, the relative errors of other samples are within 5%. The errors may be caused by different atmospheric conditions and correction methods between the experimental results of rod-plane gaps and the training samples.

The error indexes of the predicted results of different gap types under various voltage waveforms are summarized in Table 6.10. The e_{MAPE}s of rod-rod gaps under 1.2/50 μs, 1.2/5 μs, 1.5/40 μs, and 1/5 μs are respectively 1.2%, 4.6%, 0.9%, and 2.5%. For rod-plane and sphere-plane gaps, their e_{MAPE}s are 5.8 and 2.4%. The error analysis results demonstrate the validity and accuracy of the proposed method for lightning impulse discharge voltage prediction of various gap configurations.

Table 6.10 Error indexes of positive lightning impulse discharge voltage prediction results of rod-rod, rod-plane and sphere-plane gaps

Error indexes	Rod-rod gap				Rod-plane gap	Sphere-plane gap
	1.2/50 μs	1.2/5 μs	1.5/40 μs	1/5 μs	1.2/50 μs	1/50 μs
e_{MSE}	2.26	8.20	2.01	7.16	4.72	2.00
e_{MAPE}	0.012	0.046	0.009	0.025	0.058	0.024
e_{MSPE}	0.005	0.014	0.004	0.013	0.037	0.011

6.3.2 *Negative Lightning Impulse Discharge Voltage Prediction*

6.3.2.1 Experimental Data and Sample Selection

Discharge voltage test data of the IEEE standard rod-rod gaps under negative 1.2/50, 1.2/5, 1.5/40 and 1/5 μs impulse voltage waveforms [22, 23] are taken as the sample set in this section. A few data are selected as training samples, as shown in Table 6.11. The trained SVM model is used to predict the lightning impulse discharge voltages of rod-rod gaps with different gap distances and under different applied voltage waveforms. The experimental discharge voltages of the test sample are shown in Table 6.12. The data with * are caused by unstable conditions, and the average value is taken as the corresponding experimental discharge voltage. There are 7 training samples and 49 test samples, respectively. The SVM model trained by a small sample set is used to achieve the discharge voltage prediction of numerous test samples, so as to reduce the test work.

The methods to extract the electric field features and to calculate the impulse voltage waveform features are the same with those used in positive lightning impulse discharge voltage prediction, which have been introduced in Sect. 6.3.1.2.

Table 6.11 Training sample set for negative lightning impulse discharge voltage prediction

Waveform	*d* (cm)	*U* (kV)	Waveform (μs)	*d* (cm)	*U* (kV)
1.2 × 50 μs	20	176	1.2 × 5	30	274
	50	382–392*	1.5 × 40	76.2	575
	100	715	1 × 5	127	1085
	140	965	–	–	–

Table 6.12 Experimental data of the negative lightning impulse discharge voltage (kV) for test samples

d (cm)	1.2/50 μs	1.2/5 μs	*d* (cm)	1.5/40 μs	1/5 μs
20	–	188	20.32	183	188
25	217	232	22.86	203	209
30	249–260*	–	25.4	224	231
35	283–306*	316	30.48	259–271*	273
40	313–348*	358	35.56	293–318*	315
45	347–375*	405	38.1	309–342*	335
50	–	449	40.64	323–359*	360
60	455	535	45.72	360–386*	400
70	525	625	50.8	395–415*	445
80	585	710	63.5	490	550
90	670	790	76.2	–	660
100	–	880	101.6	740	875
120	835	1040	127	910	–
140	–	1120	152.4	1070	1300

6.3.2.2 Prediction Results

The improved GS algorithm is used to optimize the SVM parameters, and the optimal parameters are $C = 59.7141$, $\gamma = 0.088388$, under which the training sample set has the highest classification accuracy of 97.9592%. The trained SVM model is used to predict the discharge voltage of each test sample. The initial applied voltage range is $U_{\min} = 0$ kV, $U_{\max} = 2000$ kV, and the convergence precision is $\varepsilon = 1$ V. After iterated prediction by the golden section search method, the discharge voltage prediction results of the test samples are shown in Table 6.13.

Figure 6.21 shows the comparison of the predicted and experimental discharge voltages of the rod-rod gaps under negative lightning impulses, in which the predicted results of training samples are also plotted. It can be seen that the SVM prediction results agree well with the experimental data, and the relationships between the predicted discharge voltages and the gap distance have the same trend with those of the experimental results. The maximum relative error of the predicted results is 8.9%, while for most of the test samples it is less than 8%. It should be pointed out that the discharge voltage test results in Tables 6.11 and 6.12 are the average results obtained by different laboratories, which also have certain dispersions, and the error of some test results is up to 8%. From the perspective for engineering applications, the prediction accuracy of the above results is able to satisfy the requirements.

The error indexes of each test sample set are shown in Table 6.14. Compared with the experimental values, the e_{MAPE}s of the test samples under 1.2/50 μs, 1.2/5 μs, 1.5/40 μs and 1/5 μs lightning impulse waveforms are respectively 2.6%, 2.5%, 2.1% and 5.3%, and the e_{MAPE} for all of the 49 test samples is 3.2%. The above error indexes verify the validity and accuracy of the prediction method. It can be seen that

Table 6.13 Discharge voltage prediction results of test samples under negative lightning impulses

d (cm)	1.2/50 μs			1.2/5 μs		
	U (kV)	U_p (kV)	σ (%)	*U* (kV)	U_p (kV)	σ (%)
20	–	–	–	188	185.5	−1.3
25	217	207.0	−4.6	232	229.1	−1.3
30	249–260*	244.2	−4.1	–	–	–
35	283–306*	287.0	−2.6	316	330.6	4.6
40	313–348*	317.9	−3.8	358	366.5	2.4
45	347–375*	361.1	0.03	405	412.4	1.8
50	–	–	–	449	437.7	−2.5
60	455	469.9	3.3	535	556.5	4.0
70	525	542.5	3.3	625	644.2	3.1
80	585	587.9	0.5	710	695.5	−2.0
90	670	661.4	−1.3	790	781.0	−1.1
100	–	–	–	880	843.8	−4.1
120	835	855.8	2.5	1040	1006.8	−3.2
140	–	–	–	1120	1106.6	−1.2
d (cm)	1.5/40 μs			1/5 μs		
	U (kV)	U_p (kV)	σ (%)	*U* (kV)	U_p (kV)	σ (%)
20.32	183	180.0	−1.6	188	193.0	2.7
22.86	203	198.3	−2.3	209	217.4	4.0
25.4	224	215.1	−4.0	231	221.6	−4.1
30.48	259–271*	258.7	−2.4	273	292.7	7.2
35.56	293–318*	292.9	−4.1	315	334.3	6.1
38.1	309–342*	316.1	−2.9	335	364.9	8.9
40.64	323–359*	334.6	−1.9	360	388.5	7.9
45.72	360–386*	362.0	−3.0	400	418.9	4.7
50.8	395–415*	399.7	−1.3	445	465.9	4.7
63.5	490	500.1	2.1	550	597.4	8.6
76.2	–	–	–	660	646.4	−2.1
101.6	740	739.3	−0.1	875	890.8	1.8
127	910	909.3	−0.08	–	–	–
152.4	1070	1059.1	−1.0	1300	1219.3	−6.2

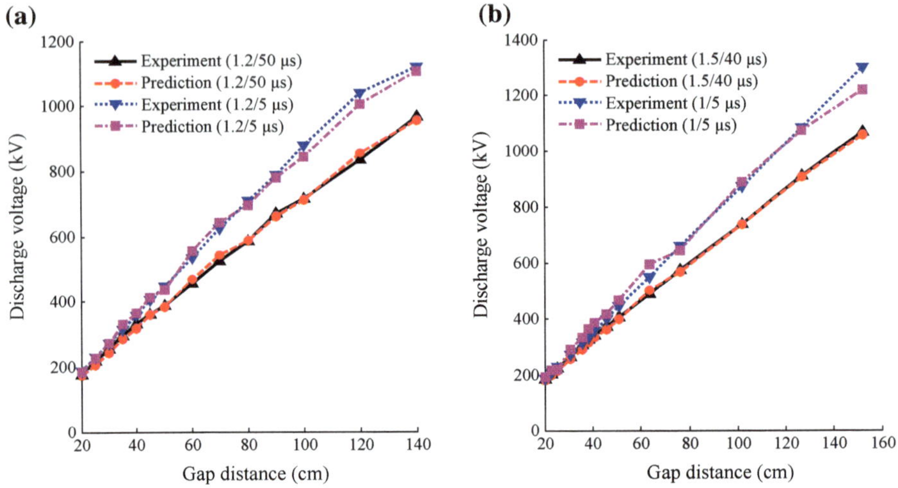

Fig. 6.21 Comparison of negative lightning impulse discharge voltage prediction results of rod-rod gaps with the experimental data: **a** 1.2/50 and 1.2/5 μs, **b** 1.5/40 and 1/5 μs

Table 6.14 Error indexes of negative lightning impulse discharge voltage prediction results of rod-rod gaps

Error indexes	Waveform			
	1.2/50 μs	1.2/5 μs	1.5/40 μs	1/5 μs
e_{MSE}	3.83	6.48	2.19	8.65
e_{MAPE}	0.026	0.025	0.021	0.053
e_{MSPE}	0.009	0.008	0.007	0.016

the discharge voltages of rod-rod gaps with different gap distances and under different impulse voltage waveforms can be predicted by the SVM model trained with a small sample data, which provides an alternative to obtain the discharge voltages instead of experiments, thus to greatly reduce the test work.

6.3.3 *Prediction of Lightning Impulse Volt-Time Characteristics*

6.3.3.1 Prediction Method

According to the idea of the air insulation prediction method, air gap breakdown is the result of the interaction of spatial features (electric field distribution) and time features (voltage waveform). The voltage waveform integral can be used to characterize the accumulation process of the impulse voltage applied to the gap. The integral of the voltage waveform from the moment the gap is applied voltage to

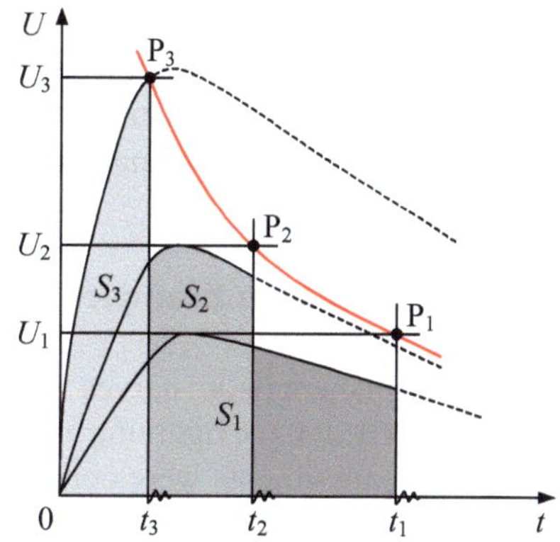

Fig. 6.22 Volt-time characteristics and voltage waveform integral

breakdown represents the energy accumulation process before gap breakdown. As shown in Fig. 6.22, the shaded part of the impulse voltage waveform is the above voltage waveform integral, which can be denoted as S_1, S_2, S_3 ….

When the breakdown occurs at the wave tail, the breakdown voltage is the voltage peak $U_{\max}$. With known front time T_f and the time to half-value T_2, the mathematical expressions of the lightning impulse voltage waveform can be obtained by solving Eq. 2.29. When the breakdown occurs at the wave front, the voltage peak $U_{\max}$ is unknown. With the known breakdown voltage U_b and the discharge time T_b, the following equation can be added to the constraint equations of Eq. 2.29:

$$u(T_0 + T_b) = A(e^{-\alpha(T_0+T_b)} - e^{-\beta(T_0+T_b)}) = U_b \tag{6.11}$$

The mathematical expressions of the lightning impulse voltage waveform can be obtained according to Eqs. 2.29 and 6.11. Then, the voltage waveform integral S_b at the breakdown moment can be calculated, which can be expressed as:

$$S_b = \int_0^{T_b} u(t)dt = \int_0^{T_b} A(e^{-\alpha t} - e^{-\beta t})dt = A\left(\frac{1 - e^{-\alpha T_b}}{\alpha} - \frac{1 - e^{-\beta T_b}}{\beta}\right) \tag{6.12}$$

Taking the breakdown voltages U_1, U_2, U_3, the discharge time t_1, t_2, t_3, and voltage waveform integrals S_1, S_2, S_3 as the input parameters of the SVM model, and several training sample data are used to train the SVM model, so as to predict the *U-t* intersections of other samples. If the gap breakdown under a given (U_x, t_x, S_x), the output of the SVM model is 1. Otherwise, the output is −1, and then the input is updated to ($U_x + \Delta U$, t_x, $S_x + \Delta S$), so repeatedly until the model outputs 1. Then $U_x + \Delta U$ is the breakdown voltage while t_x is the discharge time. Their intersection point is a point of the *U-t* curve, and $\Delta U/U_x$ is the relative error of the predicted results. Since the output of the SVM model presented in this book corresponds to the breakdown voltage, the above *U-t* characteristic curve prediction method is to determine the required voltage peak for air gap breakdown at the moment t_x under a given voltage waveshape. Moreover, in order to make the model be applicable to predict the *U-t* characteristics of different gap structures, the electric field features are also taken as the input parameters of the SVM model.

6.3.3.2 Prediction Example

The *U-t* characteristic curves of rod-rod air gaps under positive and negative 1.2/50 and 1.2/4 μs lightning impulse voltages were obtained by experiments in [27]. The experiments were conducted on vertically mounted rod-rod air gaps, one electrode is applied the high voltage and the other is grounded. The tip of the grounded rod is 1.4 m high above ground. The rod electrodes consist of 0.4 m long cylindrical brass bar with chamfered ends and the diameter is 1 cm. The gap distances d are 10 and 20 cm. Each point of the *U-t* characteristic curve was obtained by applying 5 times of lightning impulses with the same peak value. If a discharge occurred every time, the average value of the discharge time and the impulse voltage peak were respectively taken as the horizontal and vertical coordinates of the point. If one impulse did not cause gap breakdown, then 10 additional lightning impulses were applied. The tests satisfy the requirement if each of these impulses caused a discharge. In this section, the experimental data points are extracted from the *U-t* characteristic curve given in [27], as shown in Tables 6.15 and 6.16. The training samples and test samples are selected to verify the effectiveness of the SVM model for *U-t* characteristic curve prediction. The training samples are several experimental data of d = 10 cm and d = 20 cm under 1.2/50 μs standard lightning impulse voltage, and the test samples are those under 1.2/50 and 1.2/4 μs lightning impulse voltages. There are 6 training samples and 34 test samples for both of the positive and negative lightning impulse *U-t* curve prediction models.

A finite element model of rod-rod gap is established according to the experimental arrangement. The electric field distribution is calculated and the electric field features are extracted. According to Eqs. 2.29, 6.11 and 6.12, the voltage waveform integral S_b of each sample is calculated. The 28-d electric field features and (U_b, t_b, S_b) are taken as the input parameters of the SVM model, and the model is trained by the training samples shown in Tables 6.15 and 6.16. The test samples of d = 10 cm and d = 20 cm under 1.2/50 and 1.2/4 μs waveforms are taken for hybrid prediction. Given the discharge time t_b, the breakdown voltage U_b of each test sample required for gap breakdown is obtained.

The correlation analysis method is used to reduce the electric field feature dimensions. The features whose correlation coefficient with the breakdown voltage is less than 0.3 and the cross correlation coefficient is more than 0.9 are rejected. The input feature dimensions for positive and negative *U-t* curve prediction are both 11-d. The improved GS algorithm is used for parameter optimization. The optimal parameters of the positive *U-t* characteristic curve prediction model are: $C = 256$, $\gamma = 0.25$, and the error indexes of the predicted result are: $e_{SSE} = 3256.2334$, $e_{MSE} = 1.6783$, $e_{MAPE} = 0.0374$, $e_{MSPE} = 0.0085$. The optimal parameters of the negative *U-t* characteristic

Table 6.15 Experimental results of the positive *U*-*t* characteristics of rod-rod gaps (sample data): (**a**) Training sample set, (**b**) test sample set

a

Waveform (μs)	*d* (cm)	T_b (μs)	U_b (kV)	*d* (cm)	T_b (μs)	U_b (kV)
1.2/50	10	0.72	197.57	20	1.05	310.51
		1.17	143.56		2.17	206.86
		2.08	102.07		3.30	171.34

b

d (cm)	Waveform (μs)	T_b (μs)	U_b (kV)	Waveform (μs)	T_b (μs)	U_b (kV)
10	1.2/50	0.74	193.35	1.2/4	0.68	196.90
		0.77	184.57		0.75	189.75
		0.82	178.67		0.84	181.61
		0.86	172.50		0.95	166.65
		0.87	167.80		0.97	160.14
		1.01	162.44		1.04	153.32
		1.11	153.66		1.10	143.88
		1.39	132.82		1.16	133.79
		1.50	118.18		1.39	119.15
		–	–		1.59	114.43
20	1.2/50	1.14	298.43	1.2/4	1.11	276.78
		1.22	291.87		1.25	262.58
		1.31	279.79		1.33	250.68
		1.43	267.74		1.58	233.82
		1.53	255.84		1.78	220.51
		1.72	240.94		2.25	207.18
		1.82	223.54		3.00	198.31
		2.40	187.31		–	–

curve prediction model are: $C = 147.033$, $\gamma = 0.25$, and the error indexes of the predicted result are: $e_{SSE} = 1603.2663$, $e_{MSE} = 1.1777$, $e_{MAPE} = 0.0247$, $e_{MSPE} = 0.0057$.

The comparisons of the predicted and experimental *U*-*t* characteristic curves of rod-rod gaps under positive and negative lightning impulse voltages are respectively shown in Figs. 6.23 and 6.24. It can be seen that the predicted *U*-*t* characteristic curves of the 1.2/50 μs standard lightning impulse are in good agreement with the experimental results. But the predicted results of 1.2/4 μs lightning impulse have some deviations from the experimental data. That is because the training sample set is selected from the data of 1.2/50 μs lightning impulse, and the model has better prediction accuracy for the test samples with similar feature distributions with the training samples.

Table 6.16 Experimental results of the negative *U-t* characteristics of rod-rod gaps (sample data): (**a**) training sample set, (**b**) test sample set

a

Waveform	*d* (cm)	T_b (μs)	U_b (kV)	*d* (cm)	T_b (μs)	U_b (kV)
1.2/50 μs	10	0.54	208.84	20	1.14	325.24
		1.24	149.01		1.99	241.56
		2.36	107.97		3.79	188.70

b

d (cm)	Waveform (μs)	T_b (μs)	U_b (kV)	Waveform (μs)	T_b (μs)	U_b (kV)
10	1.2/50	0.59	204.14	1.2/4	0.72	204.29
		0.60	202.02		0.74	198.30
		0.66	201.39		0.79	189.68
		0.70	196.51		0.86	184.83
		0.72	191.86		0.90	176.73
		0.78	188.58		1.00	171.53
		0.80	181.11		1.02	162.12
		0.89	173.94		1.16	150.61
		1.02	167.13		1.26	137.15
		1.42	135.02		1.58	122.40
		2.17	116.75		1.80	120.46
20	1.2/50	1.21	311.77	1.2/4	1.29	276.65
		1.29	301.13		1.46	260.50
		1.42	288.18		1.76	246.50
		1.47	274.33		2.13	233.21
		1.64	259.65		2.63	225.23
		2.22	222.04		–	–
		2.64	205.40			

It can be seen from the *U-t* characteristic curve prediction results shown in Figs. 6.23 and 6.24 that:

1. When the gap breakdown takes place at the wavefront, the breakdown voltages under 1.2/50 and 1.2/4 μs waveforms are approximately the same, and their *U-t* characteristic curves almost coincide with each other, which conforms to the theoretical expectation. Since the front time of the two waveforms is the same, with the same energy applied to the gap during the wavefront, therefore the discharge time should be the same. However, the experimental results under these two impulse voltage waveforms in the wavefront stage are somewhat different, that is because the front time of the two lightning impulses is not exactly the same. It was pointed out in [27] that the front time of the standard lightning impulse was actually 1.1 μs while that of the nonstandard lightning impulse with short wave

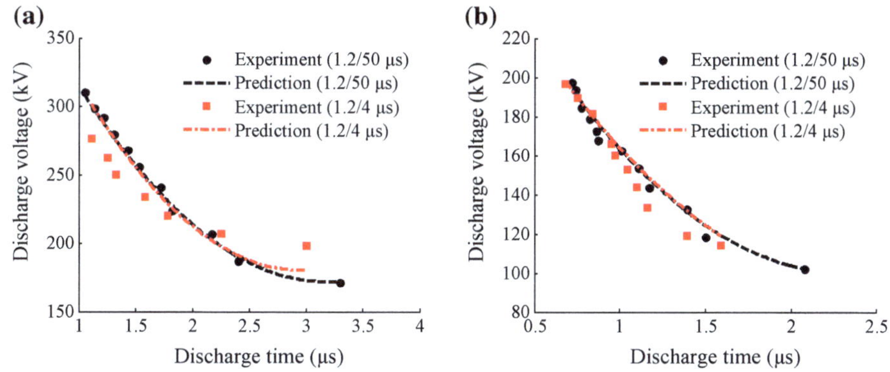

Fig. 6.23 *U-t* characteristic curve prediction of rod-rod gaps under positive lightning impulse voltages: **a** $d = 10$ cm, **b** $d = 20$ cm

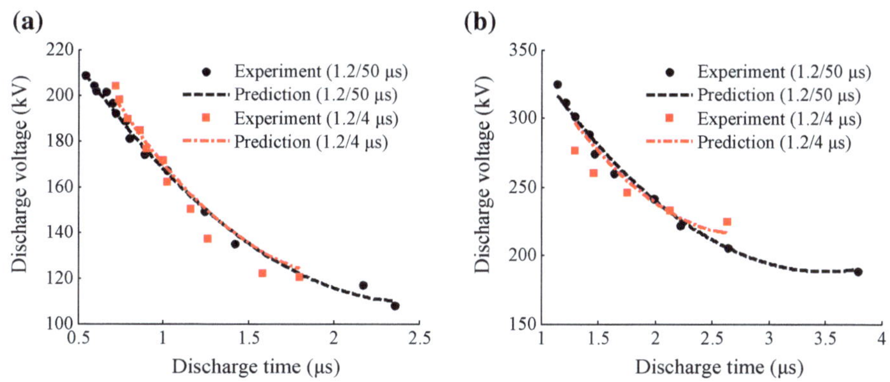

Fig. 6.24 *U-t* characteristic curve prediction of rod-rod gaps under negative lightning impulse voltages: **a** $d = 10$ cm, **b** $d = 20$ cm

tail was actually 0.9 μs. The voltage waveform features used in this prediction method are calculated by theoretical formulas, without considering the error in the experiments, and the front time is set as 1.2 μs. This is the reason leading to the deviation of the predicted and experimental *U-t* characteristic curves under 1.2/4 μs lightning impulse.

2. When the gap breakdown occurs at the wave tail, the breakdown voltages under 1.2/50 and 1.2/4 μs waveforms are different. The *U-t* characteristic curve of 1.2/4 μs waveform is above that of 1.2/50 μs, and the difference is more obvious with the longer gap. This is because the electrical stress applied on the gap under the short tail lightning impulse is less than that under the standard lightning impulse. For the short tail lightning impulse, when the impulse voltage peak increases to a sufficient extent, if the gap breakdown takes place at the same voltage peak with the standard 1.2/50 μs lightning impulse, the voltage waveform

integral S_b of the short tail 1.2/4 μs lightning impulse should be the same with the standard lightning impulse. Therefore, it requires longer discharge delay for the 1.2/4 μs lightning impulse to reach the same disruptive effect with the 1.2/50 μs lightning impulse, and thus the discharge time is longer. Or in other words, during the wave tail, if the gap breakdown under the 1.2/4 μs lightning impulse occurs at the same time with the 1.2/50 μs lightning impulse, it requires higher voltage peak to apply on the air gap.

In conclusion, the *U-t* characteristic curves predicted by the SVM model are generally consistent with the theoretical expectation and the experimental results. Trained by a few experimental data under standard lightning impulse voltage, this model is able to predict the *U-t* characteristic curve under nonstandard lightning impulse voltage, which has a certain engineering significance.

6.4 Brief Summary

In this chapter, the air insulation prediction model is used to predict the switching impulse U_{50} of rod-plane, rod-rod, rod-conductor long air gaps, the lightning impulse U_{50} and *U-t* characteristic curves of rod-rod air gaps, which can be summarized as follows:

1. The 50% discharge voltages of rod-plane and rod-rod long air gaps under positive and negative switching impulse voltages are predicted respectively, and the predicted values are in good agreement with the experimental data. The e_{MAPE}s of positive and negative rod-plane and rod-rod gaps are respectively 3.6%, 3.25%, 3.5% and 3.8%. The prediction errors of most test samples are within the standard deviation range of the experimental data, which verifies the prediction accuracy of the SVM model.
2. A hybrid model is established to predict the 50% discharge voltages of rod-plane, rod-rod and rod-conductor gaps under 20/2500 and 80/2500 μs negative switching impulse voltages. 6 experimental data of rod-plane and rod-rod gaps are taken as training samples, and the golden section search method is applied for iterated prediction. The results indicate that the e_{MAPE}s of rod-plane gaps under the switching impulses of $T_f = 20$ μs and $T_f = 80$ μs are 6.0% and 5.8% respectively, while they are 6.3% and 3.2% for rod-rod gaps. The e_{MAPE}s of the 7 test samples at the rod-conductor gap at $T_f = 80$ μs is only 1.7%. It can be seen that under certain application cases, the SVM model can be used to replace the costly and time-consuming long air gap discharge tests, thus greatly reduce the test work and cost.
3. The SVM model is used to predict the positive and negative lightning impulse discharge voltages of IEEE Std4-1995 standard rod-rod gaps under different voltage waveforms. Under the positive impulses, the predicted values are basically consistent with the experimental data. The e_{MAPE}s of the predicted results of rod-rod gaps under 1.2/50 μs, 1.2/5 μs, 1.5/40 μs, 1/5 μs impulses are respectively

1.2%, 4.6%, 0.9% and 2.5%. The e_{MAPE}s of the rod-plane and sphere-plane gaps under 1.2/50 μs and 1/50 μs are 5.8% and 2.4%, respectively. Under the negative impulses, the e_{MAPE}s of the predicted results of rod-rod gaps under 1.2/50 μs, 1.2/5 μs, 1.5/40 μs, 1/5 μs impulses are respectively 2.6%, 2.5%, 2.1% and 5.3%, and the e_{MAPE} for all of the 49 test samples is 3.2%. The predicted results demonstrate the validity of the SVM model in for lightning impulse discharge voltage prediction of air gaps.

4. The SVM model is used to predict the *U-t* characteristic curves of rod-rod gaps under positive and negative lightning impulse voltages with the waveform of 1.2/50 μs and 1.2/4 μs respectively. The predicted results are consistent with the theoretical expectation and the experimental results. Using a few experimental data of standard lightning impulse to train the SVM model, it can be used to predict the *U-t* characteristic curve under nonstandard lightning impulses, which is of a certain engineering significance.

References

1. Yan Z, Zhu DH (2007) High voltage insulation technology, 2nd edn. China Electric Power Press, Beijing
2. Wan QF, Huo F, Xie L et al (2012) Summary of research on flashover characteristics of long air-gaps. High Volt Eng 38(10):2499–2505
3. General Electric Co. (1982) Transmission line reference book-345 kV and above, 2nd edn. Electric Power Research Institute, Palo Alto, USA
4. Lings R (2005) EPRI AC transmission line reference book-200 kV and above, 3rd edn. Electric Power Research Institute, Palo Alto, USA
5. Les Renardières Group (1977) Positive discharges in long air gap discharges at Les Renardières–1975 results and conclusions. Electra 53:31–153
6. Les Renardières Group (1981) Negative discharges in long air gap discharges at Les Renardières–1978 results. Electra 74:67–216
7. Cortina R, Garbagnati E, Pigini A et al (1985) Switching impulse strength of phase-to-earth UHV external insulation-research at the 1000 kV project. IEEE Trans Power Appar Syst 104(11):3161–3168
8. Gallet G, Leroy G, Lacey R et al (1975) General expression for positive switching impulse strength valid up to extra long air gaps. IEEE Trans Power Appar Syst 94(6):1989–1993
9. Kishizima I, Matsumoto K, Watanabe Y (1984) New facilities for phase-to-phase switching impulse tests and some test results. IEEE Trans Power Appar Syst 103(6):1211–1216
10. IEC 60071-2 (1996) Insulation coordination—part 2: application guide
11. Rizk FAM (1989) A model for switching impulse leader inception and breakdown of long air gaps. IEEE Trans Power Deliv 4(1):596–606
12. Rizk FAM (1989) Switching impulse strength of air insulation: leader inception criterion. IEEE Trans Power Deliv 4(4):2187–2195
13. Carrara G, Thione L (1976) Switching surge strength of large air gaps: a physical approach. IEEE Trans Power Appar Syst 95(2):512–524
14. Chen WJ, Zeng R, He HX (2013) Research progress of long air gap discharge. High Volt Eng 39(6):1281–1295
15. Paris L (1967) Influence of air gap characteristics on line-to-ground switching surge strength. IEEE Trans Power Appar Syst 86(8):936–947
16. Paris L, Cortina R (1968) Switching and lightning impulse discharge characteristics of large air gaps and long insulator strings. IEEE Trans Power Appar Syst 87(4):947–957

17. Thione L, Pigini A, Allen NL (1992) Guidelines for the evaluation of the dielectric strength of external insulation. CIGRE Brochure, Paris, France
18. Watanabe Y (1967) Switching surge flashover characteristics of extremely long air gaps. IEEE Trans Power Appar Syst 86(8):933–936
19. Qiu ZB, Ruan JJ, Tang LZ et al (2018) Energy storage features and discharge voltage prediction of air gaps. Trans China Electrotech Soc 33(1):185–194
20. Wang Y, Wen XS, Lan L et al (2014) Breakdown characteristics of long air gap with negative polarity switching impulse. IEEE Trans Dielectr Electr Insul 21(2):603–611
21. Qiu ZB, Ruan JJ, Xu WJ et al (2017) Energy storage features and a predictive model for switching impulse flashover voltages of long air gaps. IEEE Trans Dielectr Electr Insul 24(5):2703–2711
22. The subcommittee on correlation of laboratory data of EEI-NEMA joint committee on insulation co-ordination (1937) Flashover characteristics of rod gaps and insulators. Trans AIEE 56(6): 712–714
23. IEEE Std 4-1995. IEEE standard techniques for high-voltage testing
24. Abdullah M, Kuffel E (1965) Development of spark discharge in nonuniform field gaps under impulse voltages. Proc IEE 112(5):1018–1024
25. Mavroidis PN, Mikropoulos PN, Stassinopoulos CA (2007) Discharge characteristics in short rod-plane gaps under lightning impulse voltages of both polarities. In: Paper presented at the 42nd international universities power engineering conference, Brighton, UK, 4–6 September 2007
26. Qiu ZB, Ruan JJ, Huang CP et al (2018) A numerical approach for lightning impulse flashover voltage prediction of typical air gaps. J Electr Eng Technol 13(3):1326–1336
27. Ancajima A, Carrus A, Cinieri E et al (2007) Breakdown characteristics of air spark-gaps stressed by standard and short-tail lightning impulses: experimental results and comparison with time to sparkover models. J Electrostat 65(5–6):282–288

Chapter 7
Engineering Applications of Air Insulation Prediction Model

7.1 Discharge Voltage Prediction of Parallel Gaps for Transmission Line Insulator String

The parallel gap for insulator string is used as an air gap device for lightning protection of overhead transmission line, which is a typical application of air insulation in power transmission project. The parallel gap is usually composed of metal electrodes and connecting hardware installed at the high voltage side and grounded side of the insulator string. In recent years, much research work has been carried out on the lightning impulse characteristics, insulation coordination and arc-guiding performance of parallel gaps with different structures, which accumulate a lot of useful experimental data. Limited by test conditions, the experimental lightning impulse 50% discharge voltages of the parallel gaps for 220 kV transmission line insulator strings given in [1, 2] are taken for comparisons, the U_{50} of these parallel gaps are predicted by the proposed SVM model, so as to verify its applicability in discharge voltage prediction of engineering gaps.

7.1.1 Experimental Data and Samples

The discharge tests of parallel gaps presented in [1, 2] were carried out in the outdoor testing ground of the National Engineering Laboratory for Ultra High Voltage Engineering in Kunming, China. The applied voltage waveform is the negative 1.2/50 μs standard lightning impulse voltage waveform. The insulators involved in the test include XP-70 porcelain insulator (14 piece string) and FXBW4-220/100 composite insulator. The electrodes of the parallel gaps include rod-shaped electrodes and ring-shaped electrodes. For the porcelain insulator string, the gap distances of the rod-shaped parallel gaps are 1.722 and 1.836 m. For the composite insulator, the gap distances of the rod-shaped parallel gaps are 1.744, 1.844 and 1.944 m, while those of the ring-shaped parallel gaps are 1.644, 1.744 and 1.844 m.

Z. Qiu et al., *Air Insulation Prediction Theory and Applications*, Power Systems,
https://doi.org/10.1007/978-981-10-5163-0_7

Table 7.1 Experimental U_{50} of parallel gaps for 220 kV insulator string

Insulator types	Parallel gaps (m)	U_{50t} (kV)	U_{50c1} (kV)	U_{50c2} (kV)	U_{50c3} (kV)	σ (%)
Porcelain insulator	Rod-1.722	938.3	1214.1	1187.7	1157.0	2.65
	Rod-1.836	1049.6	1358.1	1328.6	1294.2	2.25
Composite insulator	Rod-1.744	793.2	1026.3	1004.1	978.1	4.68
	Rod-1.844	876.4	1134.0	1109.4	1080.6	3.65
	Rod-1.944	966.5	1250.6	1223.4	1191.7	3.25
Composite insulator	Ring-1.644	852.0	1102.4	1078.5	1050.6	–
	Ring-1.744	891.0	1152.9	1127.9	1098.6	–
	Ring-1.844	954.4	1234.9	1208.1	1176.8	–

The test site locates at a high altitude area, with the altitude of 2100 m, but the training samples used for prediction are experimental discharge voltages of rod-rod gaps at zero altitude area. Therefore, it is necessary to correct the experimental data in [1, 2] to zero altitude conditions according to the altitude correction formulas, so as to be used for comparison with the discharge voltage prediction results. Many altitude correction formulas for air gap discharge voltage have been proposed, in which the formulas to correct the lightning impulse discharge voltage given by IEC 60071-1: 2006 [3] and [4, 5] are respectively.

$$K_{a1} = e^{\frac{H}{8150}} \tag{7.1}$$

$$K_{a2} = \frac{1}{1 - 10^{-4} \cdot H} \tag{7.2}$$

$$K_{a3} = \frac{1}{1 - 10^{-4} \cdot 0.9H} \tag{7.3}$$

where H is the altitude in m. The experimental U_{50} of the parallel gaps for 220 kV insulator string and the standard deviations obtained in high altitude area were given in [1, 2], which are shown as U_{50t} and σ in Table 7.1. The experimental results are corrected to zero altitude conditions using Eqs. 7.1–7.3 respectively, which are shown as U_{50c1}, U_{50c2}, and U_{50c3} in Table 7.1. The above values will be used for comparison with the SVM prediction results.

7.1.2 *Electric Field Calculation*

According to the electrode structure parameters of the parallel gaps and the test configurations given in [1, 2], the three-dimensional simulation models are established by SolidWorks software, as shown in Fig. 7.1. The rod electrodes or the ring elec-

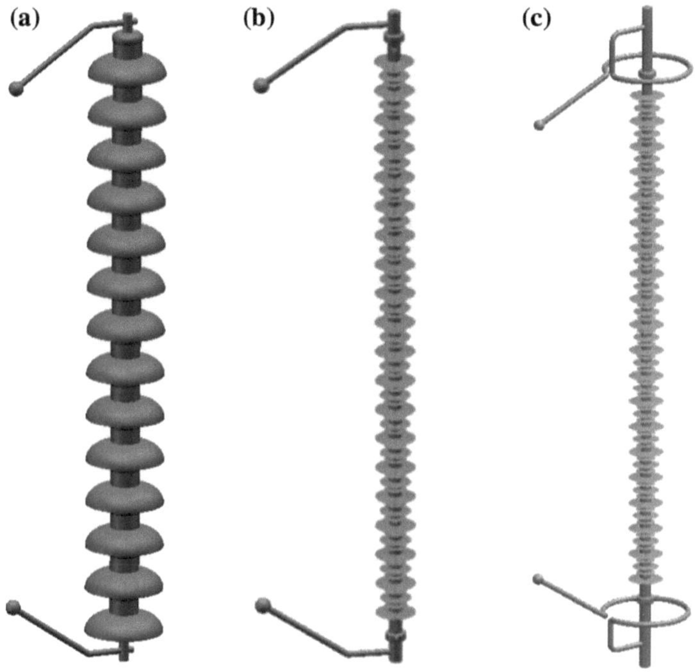

Fig. 7.1 Simulation models of parallel gaps for transmission line insulator string: **a** rod-shaped parallel gaps of porcelain insulator string, **b** rod-shaped parallel gaps of composite insulator, **c** ring-shaped parallel gaps of composite insulator

trodes of the parallel gaps are installed at both ends of the insulator. The structure parameters of the electrodes were depicted in [2]. For the rod-shaped parallel gaps, the length of the rod electrode is 35 cm and the rod tip is a spherical head with the diameter of 5 cm. For the ring-shaped parallel gaps, the ring electrode is with an extended rod, and mounted at the insulator end with a L-shaped connecting rod. The ring electrode is an open ring with the inner diameter of 30 cm, and the opening spacing is 2 cm. The length of the extended rod electrode is 30 cm and the rod tip is a spherical head with the diameter of 4 cm. The vertical height of the L-shaped connecting rod is 5 cm, and the horizontal length is 15 cm.

The above models are imported into a finite element analysis software to calculate the electric field distribution. The shortest discharge path is defined as the shortest straight path between the high voltage electrode and the grounded electrode. The discharge channel is defined as a cylindrical region between two electrodes, with the diameter of the spherical head. The high voltage fittings of the insulator and the high voltage electrode of the parallel gap are applied unit potential 1 V, and the grounded fittings of the insulator and the grounded electrode of the parallel gap are applied zero potential. Taking 1.722 m rod-shaped parallel gap of porcelain insulator, 1.744 m rod-shaped parallel gap and 1.744 m ring-shaped parallel gap of composite insulator

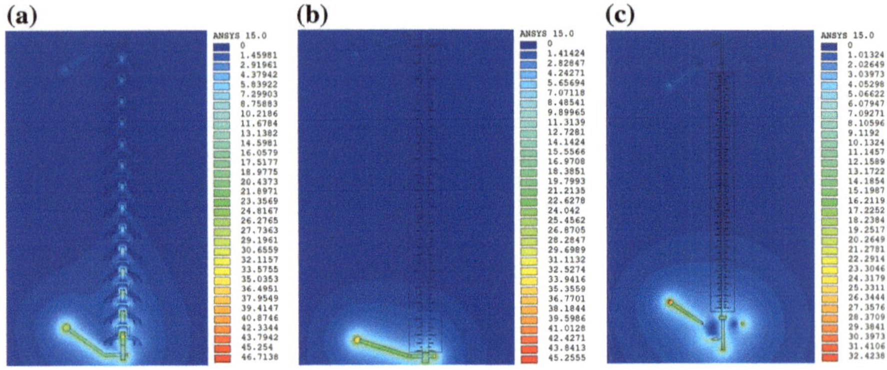

Fig. 7.2 Electric field distributions of parallel gaps for transmission line insulator string: **a** rod-shaped parallel gaps of porcelain insulator string, **b** rod-shaped parallel gaps of composite insulator, **c** ring-shaped parallel gaps of composite insulator

as examples, their electric field distributions are shown in Fig. 7.2. It can be seen that the electric field is concentrated near the high voltage electrode of the parallel gap, and the field strength at the electrode tip is higher than that of the fittings at the insulator end. The electric field strength gradually decreases along the shortest path. Besides, there is a certain electric field concentration near the grounded electrode, but the field strength is relatively small. The maximum field strength values of the rod-shaped parallel gaps for porcelain insulator and composite insulator with similar gap distances are very close. For the rod-shaped and ring-shaped parallel gaps for composite insulator with the same gap distance, the maximum field strength of the ring-shaped gap is less than that of the rod-shaped gap. This is because the ring electrodes make the electric field distributions more uniform at two ends of the insulator.

7.1.3 Prediction Results and Analysis

The experimental values of negative 1.2/50 μs lightning impulse discharge voltage of rod-rod gaps with the distances ranging from 1 to 2 m are selected as training samples, which were given in IEEE Std4-1995. The arrangement of the rod-rod gap is shown in Fig. 6.17, and the discharge voltages are shown in Table 7.2. The trained SVM model is used to predict the discharge voltages of the above 3 types of parallel gaps, and the predicted results are shown in Table 7.2, where $U_{50\mathrm{p}}$ is the predicted values, σ_1, σ_2 and σ_3 are the relative errors of the predicted values compared with $U_{50\mathrm{c}1}$, $U_{50\mathrm{c}2}$ and $U_{50\mathrm{c}3}$.

It can be seen from Table 7.2 that the predicted U_{50} of the parallel gaps are close to the experimental values. Compared to the altitude correction results obtained by 3 formulas, the e_{MAPE}s of the 8 test samples are respectively 3.03%, 3.55% and

Table 7.2 Comparison of predicted and experimental U_{50} of parallel gaps for 220 kV insulator string

Training samples		Predicted results					
d (m)	U_{50} (kV)	Insulator types	Parallel gaps (m)	U_{50p} (kV)	σ_1 (%)	σ_2 (%)	σ_3 (%)
1.0	715	Porcelain insulator	Rod-1.722	1210.9	−0.3	1.9	4.7
1.2	835		Rod-1.836	1275.0	−6.1	−4.0	−1.5
1.4	965	Composite insulator	Rod-1.744	1036.3	1.0	3.2	6.0
1.6	1090		Rod-1.844	1097.3	−3.2	−1.1	1.5
1.8	1240		Rod-1.944	1145.9	−8.4	−6.3	−3.8
2.0	1340	Composite insulator	Ring-1.644	1114.5	1.1	3.3	6.1
–	–		Ring-1.744	1118.5	3.1	5.4	8.2
–	–		Ring-1.844	1246.7	1.0	3.2	5.9

4.71%. The maximum absolute percentage errors are respectively 8.4%, 6.3% and 8.2%. There are some differences between the electric field simulation models and the actual test configurations, results in deviations of the calculated electric field features and the electric field distributions under test conditions, which is one of the reasons leading to the prediction errors. In general, the errors are within the acceptable range for engineering applications. Using 6 experimental data of rod-rod gap discharge voltage to train the SVM model, it is able to predict the lightning impulse U_{50} of the parallel gaps, which verify the validity of the prediction method.

According to the above discharge voltage prediction results of the 3 parallel gap structures, the relations between the U_{50} and the gap distance d can be fitted, as shown in Fig. 7.3. For the composite insulator, under the same gap distance, the U_{50} of the ring-shaped parallel gap is higher than that of the rod-shaped parallel gap. This is because the ring electrodes make the electric field distribution more uniform at two ends of the insulator, and therefore the electric field distribution of the ring-shaped parallel gap is more uniform than that of the rod-shaped parallel gap. Under the close gap distance, the U_{50} of the parallel gap for porcelain insulator is higher than that of the parallel gap for composite insulator. The suspended electrodes like the iron cap, steel foot in porcelain insulator string make its electric field distribution more uniform than that of the composite insulator. In addition, under close gap distances, the U_{50} of the rod-shaped parallel gaps for porcelain insulator is higher than that of the ring-shaped parallel gaps for composite insulator. The above conclusions are consistent with the laws of the experimental results in [1, 2], which further verifies the rationality of the SVM prediction results.

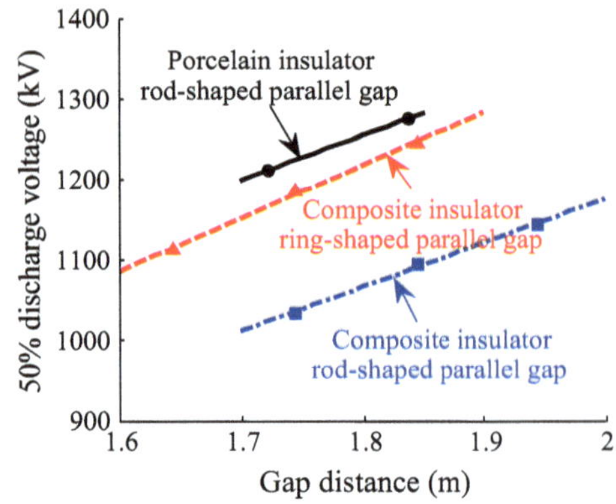

Fig. 7.3 Relationship between the predicted U_{50} and the gap distances of parallel gaps

7.2 Discharge Voltage Prediction of Transmission Line-Tower Gaps

Discharge voltage prediction of practical air gaps in transmission projects is a long-sought goal and also a great challenge in high voltage engineering. In this section, three types of transmission line-tower air gaps are selected as the research objects, including the middle-phase air gaps of the 750 kV double-circuit tower, the bottom-phase V-string air gaps of the UHV compact tower, and the side-phase air gaps of the UHV cup-type tower. Together with rod-plane long air gaps, these engineering gaps are taken as the sample set, and their finite element simulation models are established for electric field calculation and feature extraction. Then the SVM model is used to predict the standard switching impulse discharge voltages of these engineering gaps, and the results are compared with the experimental data in references.

7.2.1 *Transmission Line-Tower Air Gaps*

7.2.1.1 Middle-Phase Air Gaps of 750 kV Double-Circuit Tower

The middle-phase air gap of the 750 kV double-circuit tower is shown in Fig. 7.4 [6]. The tower is a 1:1 full-scale simulated tower head, and the conductor is a 1:1 scale simulated bundled conductor made of galvanized iron tube with the diameter of about 28 mm. The subconductor spacing is 400 mm the length is about 20 m. A grading ring with the diameter of 1.5 m is equipped at both ends of the conductor. In order to hang the simulated tower head on the 60 m high gantry tower, the middle-phase air gaps of the simulated tower head are taken for discharge tests separately with the bottom phase. The height of the conductor above ground is greater than or equal to 20 m.

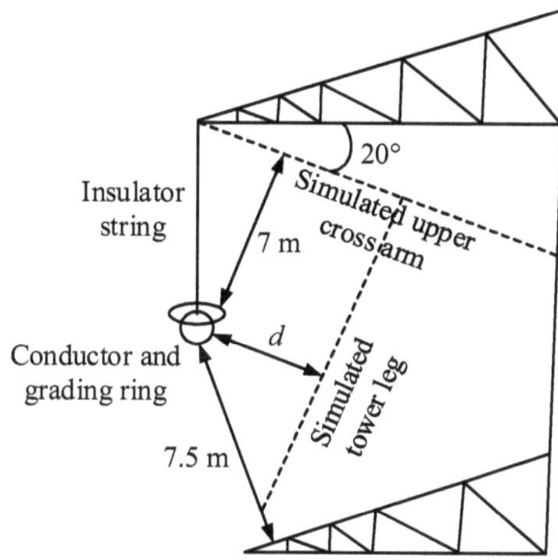

Fig. 7.4 Schematic diagram of the middle-phase air gap on 750 kV double-circuit tower

The switching impulse discharge tests of the middle-phase air gaps on the 750 kV double-circuit tower were conducted in [6]. The applied voltage is the positive 250/2500 μs standard switching impulse voltage. In actual situation, the suspension insulator will produce a windage yaw angle α under strong wind. The angle α is determined by the wind speed, wind direction, insulator and conductor structure. It is difficult to simulate windage yaw condition by pulling the insulator string, therefore, the simulation of α is realized by rotating the simulated tower structure, shown as the dotted line in Fig. 7.4. For the middle-phase air gap, the windage yaw angle α is 20°. The minimum gap distance d between the conductor (including the grading ring) and the simulated tower leg is the selected experimental gap distance, and d ranges from 4.0 to 6.4 m.

7.2.1.2 Bottom-Phase V-String Air Gaps of the UHV Compact Tower

The bottom-phase V-string air gap of the UHV compact tower is shown in Fig. 7.5 [7]. The tower is a self-supporting single pylon tower. The simulated transmission line is a 10-bundled 500 mm^2 conductor and the bundle spacing is 400 mm. The 10-bundled conductor is made of stainless steel tube, with the length of 20 m, and the diameter of the steel tube is 30 mm. The diameter of the bundled circle is 1.29 m, and the suspension angle of the conductor is 10°, since the conductor will droop due to its weight and thus to form an angle with the horizontal plane. The V-string and the middle I-string (the middle suspended string of the bottom-phase V-string) are assembled by the 9.75 m long 1000 kV composite insulators and the fittings.

The switching impulse discharge tests of the bottom-phase V-string air gaps of the UHV compact tower were conducted in [7]. As shown in Fig. 7.5, the phase spacings between A, B and C-phase are adjusted to 15 m. The distance d between the conductor and the framework ranges from 5.0 to 7.4 m. The bottom-phase (B-phase)

Fig. 7.5 Schematic diagram of the bottom-phase V-string air gap of UHV compact tower

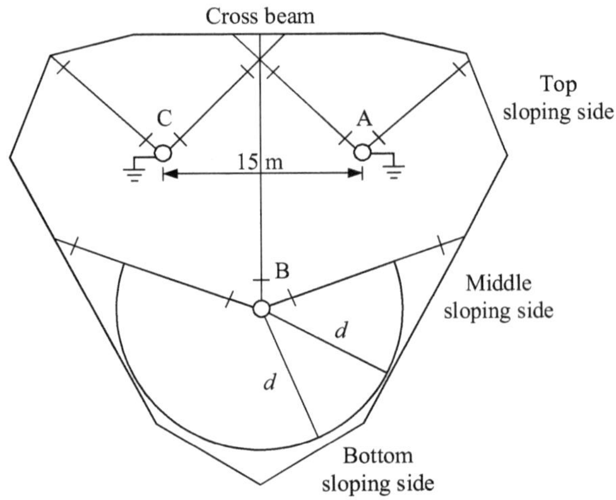

conductor is applied positive 250/2500 μs standard switching impulse voltage. The upper-phase (A, C-phase) conductors are grounded respectively. At the position near the bottom-phase high voltage electrode in the tower window, the minimum distance between the middle sloping side and the grading ring is 6.5 m and the minimum distance between the bottom sloping side and the conductor is 8.5 m.

7.2.1.3 Side-Phase Air Gaps of UHV Cup-Type Tower

The side-phase air gap of the UHV cup-type tower is shown in Fig. 7.6 [8]. The cup-type tower is manufactured according to the design drawing of ZBS2-type tower. The tower height is 71 m, the width is 57 m, and the length of the cross arm is 13.4 m. The I-type insulator is suspended on the side phase. The conductor is a simulated hard conductor wire machined with 1:1 scale of 8-bundled 500 mm^2 conductor. The sub conductor is a 20 m long galvanized steel tube with the diameter of about 30 mm. The sub conductor interval is 400 mm, and a shielding ring with the diameter of 1.5 m is installed at both ends of the conductor. The horizontal contained angle of the bundled conductor is 10°. The suspension insulator is a composite insulator used in the 1000 kV UHV transmission line. The outer diameter of the grading ring is 1.1 m, and the pipe diameter is 100 mm.

The switching impulse discharge tests of the side-phase air gaps on the UHV cup-type tower were conducted in [8]. The tests were carried out at the outdoor testing ground of State Grid Electric Power Research Institute in Wuhan, China. The test power supply is a 5400 kV impulse voltage generator equipped with relevant measuring system, which can produce 250/2500 μs standard switching impulse voltage. Considering the windage yaw condition of I-type insulator, an insulated rope is used to pull the bundled conductor, so that a 23° windage yaw angle will form between the insulator string and the vertical plane. The distance between the grading ring and

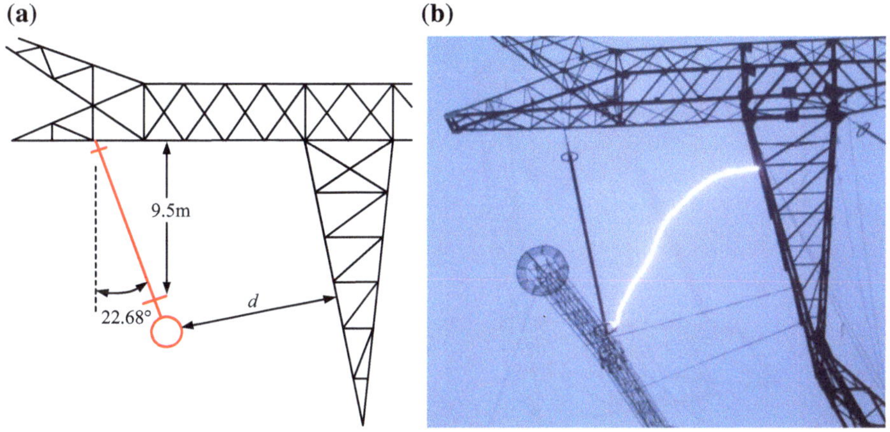

Fig. 7.6 Schematic diagram of the side-phase air gap on the UHV cup-type tower: **a** schematic diagram, **b** test arrangement

Table 7.3 Experimental U_{50} of transmission line-tower air gaps under standard switching impulse voltage

Middle-phase air gaps of 750 kV double-circuit tower		Bottom-phase V-string air gaps of UHV compact tower		Side-phase air gaps of UHV cup-type tower	
d (m)	U_{50} (kV)	*d* (m)	U_{50} (kV)	*d* (m)	U_{50} (kV)
4.0	1421	5.0	1681	4.5	1577
4.5	1549	6.0	1802	5.5	1803
5.0	1677	7.0	1896	6.4	1971
5.5	1804	7.4	1977	7.5	2115
6.0	1855	–	–	8.0	2240
6.4	1893	–	–	–	–

the cross arm is 9.5 m. The test configuration is shown in Fig. 7.6b. By adjusting the suspension position of the insulator string on the cross arm, the gap distance *d* between the side-phase conductor and the tower body can be adjusted to 4.5 m, 5.5 m, 6.4 m, 7.5 m and 8.0 m respectively. The simulated 8-bundled conductor is applied a positive 250/2500 μs standard switching impulse voltage. The U_{50} of the conductor-tower air gaps with different gap distances were measured by up-and-down method, while each gap was applied 40 times of switching impulse.

The experimental 50% discharge voltages of the above 3 types of transmission line-tower air gaps under standard switching impulse voltage are shown in Table 7.3 [6–8]. These engineering gaps are taken as a sample set to verify the effectiveness of the SVM model for discharge voltage prediction of the transmission line-tower air gaps.

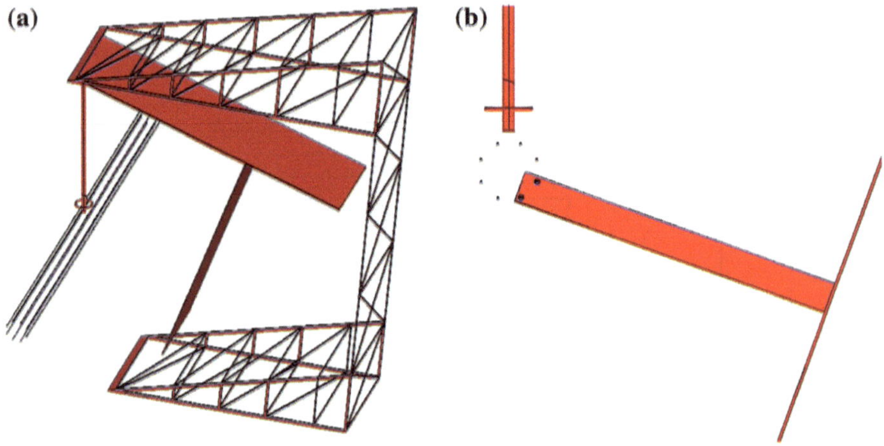

Fig. 7.7 Simulation model of the middle-phase air gap on 750 kV double-circuit tower: **a** the simulation model, **b** the discharge channel

7.2.2 *Electric Field Calculation and Feature Extraction*

7.2.2.1 Middle-Phase Air Gaps of 750 kV Double-Circuit Tower

According to the test configuration shown in Fig. 7.4 and the corresponding structure parameters [6], the simulation model of the middle-phase air gap on the 750 kV double-circuit tower is established, as shown in Fig. 7.7a. The height of the conductor above ground is 22 m, and the insulator is simplified by the cylinder. The simulated upper cross arm and the tower leg are replaced by metal plates with the width of 4 m.

The conductor and its nearby grading ring are applied 1 V, and the tower window is applied zero potential. The size of outer air layer is 80 m × 60 m × 60 m, and the outer air boundary is applied zero potential. Taking the air gap with $d = 5.5$ m for example, the potential and electric field distributions of the middle-phase air gap on the 750 kV double-circuit tower are shown in Fig. 7.8. The electric field distributions of the air gaps with the distances of 4.0, 4.5, 5.0, 5.5, 6.0 and 6.4 m are calculated successively. The discharge channel is shown in Fig. 7.7b. The shortest discharge path is the path of the minimum distance between the sub-conductor and the simulated tower leg. The electric field features in the discharge channel and on the shortest discharge path are extracted for discharge voltage prediction.

7.2.2.2 Bottom-Phase V-String Air Gaps of UHV Compact Tower

According to the air gap configuration of the UHV compact tower [7] shown in Fig. 7.5, the finite element model is established on the basis of some reasonable simplifications, as shown in Fig. 7.9a. In actual conditions, the V-type insulator

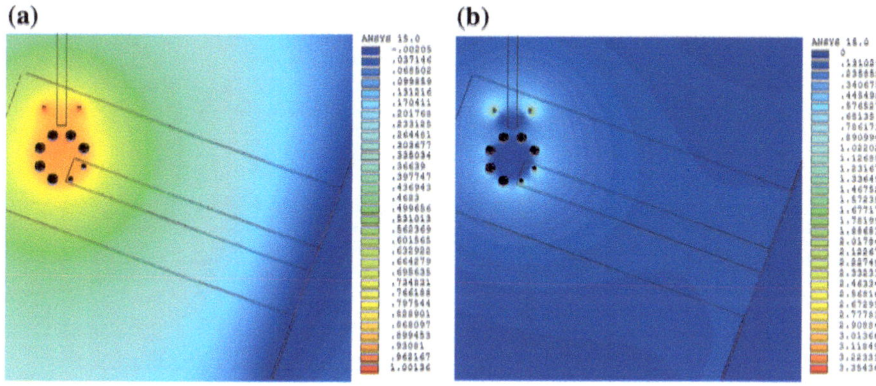

Fig. 7.8 Potential and electric field distribution of the middle-phase air gap on the 750 kV double-circuit tower ($d = 5.5$ m): **a** potential distribution, **b** electric field distribution

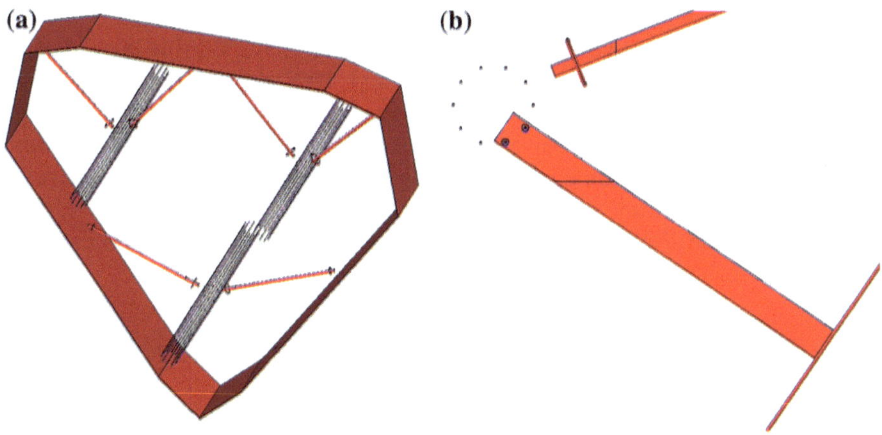

Fig. 7.9 Simulation model of UHV compact transmission line air gap: **a** the whole model, **b** the discharge channel

string used to suspend the bottom-phase conductor may droop naturally, and the suspension angle of the conductor is about 10°. Considering that the influence of the insulator on the electric field distribution is limited, the insulator is simplified and replaced by a long straight cylinder, while the insulator sheds are neglected. The tower window in actual test arrangement is well grounded by a soft copper strip, in the simulation model the tower window is replaced by a thin metal plate with the width of 5 m.

The bottom-phase V-string conductor and the nearby grading ring are applied 1 V, the upper-phase conductors and their nearby grading rings, the tower window and the grounded grading rings are applied zero potential. The outer air layer in the simulation is with the size of 80 m × 60 m × 60 m, while the outer air boundary is also applied zero potential. Taking the air gap with $d = 7.0$ m for example, the

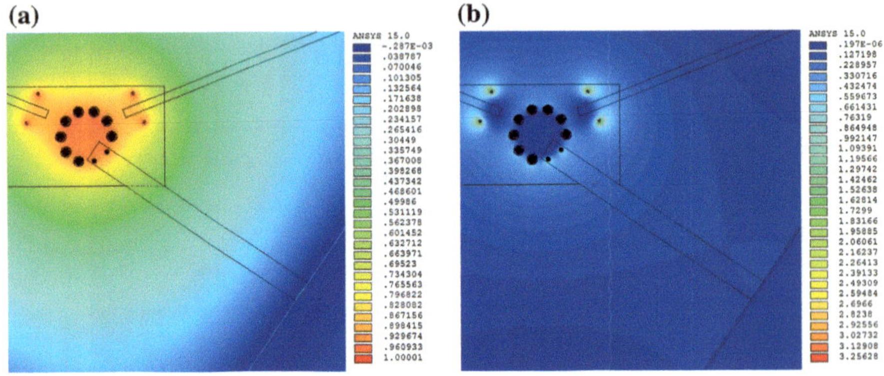

Fig. 7.10 Potential and electric field distributions of the bottom-phase V-string air gap of the UHV compact tower ($d = 7$ m): **a** potential distribution, **b** electric field distribution

potential and electric field distributions of the bottom-phase V-string air gap of the UHV compact tower are shown in Fig. 7.10.

The electric field distributions of the air gaps with the distances of 5.0, 6.0, 7.0 and 7.4 m are calculated successively. The discharge channel is defined as shown in Fig. 7.9b. The shortest discharge path is the path of the minimum distance between the sub-conductor and the middle slopping side of the tower window. The electric field features in the discharge channel and on the shortest discharge path are extracted for discharge voltage prediction.

7.2.2.3 Side-Phase Air Gaps of UHV Cup-Type Tower

According to the experimental arrangement shown in Fig. 7.6 and the structure parameters [8] of the full-scale UHV cup-type tower, conductor, insulator and fittings, a three-dimensional simulation model of the side-phase air gap is established by SolidWorks software, as shown in Fig. 7.11.

In order to extract the electric field features, the discharge channel and the shortest discharge path of the conductor-tower air gap are defined as shown in Fig. 7.12. The discharge path is the straight path from the nearest sub-conductor to the tower body, and the discharge channel is a spatial region between the conductor and the tower, which contains two sub-conductors. It should be pointed out that the discharge path is with large dispersion in actual discharge tests. According to the experimental statistics, in most cases, the discharge is from the conductor or the grading ring to the tower. The discharge may start from the conductor, and may also start from the grading ring, as shown in Fig. 7.6b. For the above three transmission line-tower air gaps, the discharge propagation is not always along the shortest path between the conductor and the tower. In the electric field simulations and the following prediction

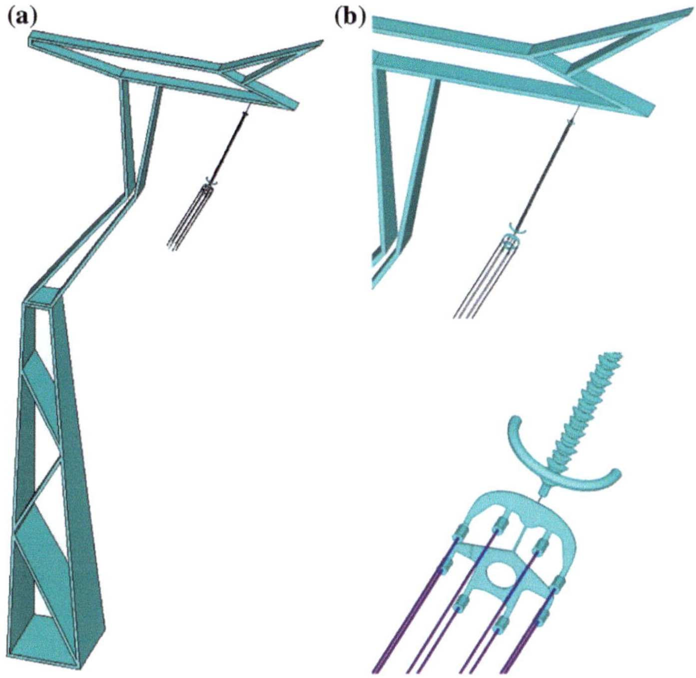

Fig. 7.11 Simulation model of the side-phase air gap on the UHV cup-type tower: **a** whole model, **b** partial enlarged detail

Fig. 7.12 Schematic diagram of the discharge channel and the shortest discharge path of the conductor-tower air gap

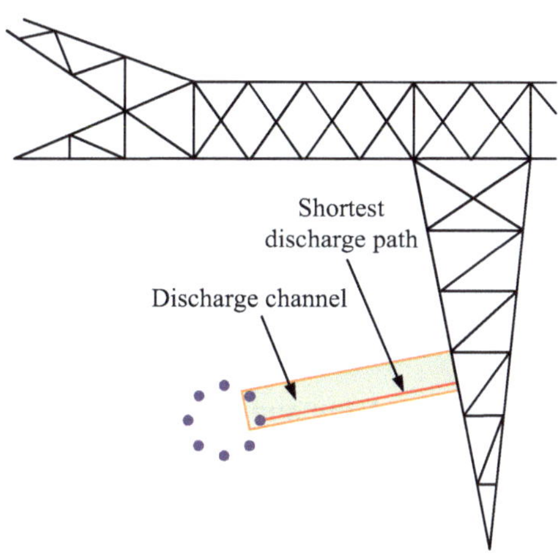

studies, only the ideal conditions are considered without regard to the dispersion of the discharge path.

The regions near the bundled conductor and the grading ring are meshed more intensively. The bundled conductor and high voltage fittings are applied unit potential

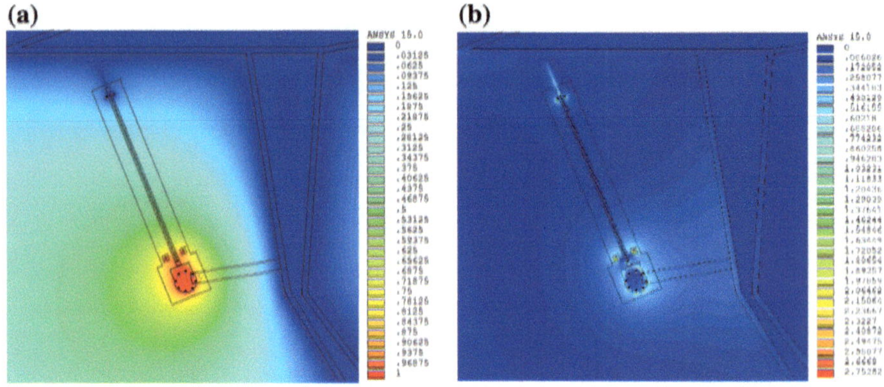

Fig. 7.13 Potential and electric field distribution of the side-phase air gap on the UHV cup-type tower: **a** potential distribution, **b** electric field distribution

1 V, and the tower, the grounded fittings and the outer air boundary are applied zero potential. Taking the air gap with $d = 4.5$ m for example, the potential and electric field distributions of the side-phase air gap of the UHV cup-type tower are shown in Fig. 7.13.

It can be seen from Fig. 7.13 that the electric field is concentrated near the bundled conductor and the grading ring, and the electric field strength gradually decreases along the path from the bundled conductor to the tower. The discharge channel defined in Fig. 7.12 contains the area where the electric field strength gradually decreases, therefore the electric field features extracted from this area can be used to characterize the electric field distribution of the conductor-tower air gap. The electric field distributions of the conductor- tower air gaps with the gap distance of 4.5, 5.5, 6.4, 7.5 and 8.0 m are calculated successively, and their electric field features are extracted for discharge voltage prediction.

7.2.3 Discharge Voltage Prediction

7.2.3.1 Training and Test Sample Set

A reasonable training sample set is very important to guarantee the generalization performance of the SVM model. On the basis of the fact that the gap factor method has been widely accepted in high voltage engineering to evaluate the discharge characteristics of engineering gap configurations, here, the experimental data of rod-plane air gaps are selected as a part of the training sample set.

The discharge tests of the long rod-plane air gaps were carried out in [9], under the positive 120/4000 μs switching impulse voltage. According to the experimental

Table 7.4 50% discharge voltages of rod-plane air gaps

d (m)	U_{50} (kV)	d (m)	U_{50} (kV)
2.0	759	5.0	1320
2.5	867	5.5	1390
3.0	969	6.0	1470
3.5	1062	6.5	1540
4.0	1150	7.0	1610
4.5	1240	–	–

results, the empirical formula to calculate the U_{50} under different gap distances was put forward, which can be expressed as:

$$U_{50} = 500 \cdot d^{0.6} \tag{7.4}$$

CIGRE Brochure [10] and IEC 60071-2 [11] consider that Eq. 7.4 can give a better approximation for the discharge voltages of rod-plane gaps under 250/2500 μs standard switching impulse. The U_{50} of rod-plane air gaps with the gap distance ranging from 2.0 to 7.0 m are shown in Table 7.4 [9].

The electrode structure parameters of the rod-plane gap were given in [9]. The rod electrode is a vertically-arranged square rod with 1 cm^2 section. The plane electrode is a grounded metal plate on the floor. According to the above parameters, the three-dimensional finite element models of the rod-plane gaps with different distances are established. Their electric field distributions are calculated for feature extraction, and taken to train the SVM model.

In addition, in order to reflect the influence of electric field variations on air-gap discharge characteristics, one type of engineering gap configuration should be selected as the training sample. This can be seen as that the SVM model will learn from these data how to determine the value of the gap factor k for different gap configurations. Therefore, except for the rod-plane gaps, one of the above three kinds of EHV and UHV transmission line-tower air gaps is successively taken as another part of the training sample set, and the other two engineering configurations are taken as the test samples. The training sample sets containing the middle-phase air gaps of 750 kV double-circuit tower, the bottom-phase V-string air gaps of UHV compact tower, and the side-phase air gaps of UHV cup-type tower are respectively denoted as A, B and C.

7.2.3.2 Prediction Results and Analysis

The improved GS algorithm is applied for parameter optimization of the SVM model based on 5-fold cross validation. The input parameters are 28-d electric field features. The golden section search method is used for the iterated prediction. The initial search interval is set as 0–3000 kV and the convergence precision is 5 kV. The

Table 7.5 U_{50} prediction results of transmission line-tower air gaps with training sample set A

Bottom-phase V-string air gaps of UHV compact tower				Side-phase air gaps of UHV cup-type tower			
d (m)	U_{50} (kV)	U_{50p} (kV)	σ (%)	d (m)	U_{50} (kV)	U_{50p} (kV)	σ (%)
5.0	1681	1759	4.64	4.5	1577	1425	−9.64
6.0	1802	1954	8.44	5.5	1803	1655	−8.31
7.0	1896	2101	10.81	6.4	1971	1827	−7.45
7.4	1977	2160	9.26	7.5	2115	2097	−0.90
–	–	–	–	8.0	2240	2194	−2.14

Table 7.6 U_{50} prediction results of transmission line-tower air gaps with training sample set B

Middle-phase air gaps of 750 kV double-circuit tower				Side-phase air gaps of UHV cup-type tower			
d (m)	U_{50} (kV)	U_{50p} (kV)	σ (%)	d (m)	U_{50} (kV)	U_{50p} (kV)	σ (%)
4.0	1421	1541	8.4	4.5	1577	1648	4.5
4.5	1549	1618	4.5	5.5	1803	1915	6.2
5.0	1677	1691	0.8	6.4	1971	2062	4.6
5.5	1804	1744	−3.3	7.5	2115	2216	4.8
6.0	1855	1786	−3.7	8.0	2240	2258	0.8
6.4	1893	1801	−4.9	–	–	–	–

Table 7.7 U_{50} prediction results of transmission line-tower air gaps with training sample set C

Middle-phase air gaps of 750 kV double-circuit tower				Bottom-phase air gaps of UHV compact tower			
d (m)	U_{50} (kV)	U_{50p} (kV)	σ (%)	d (m)	U_{50} (kV)	U_{50p} (kV)	σ (%)
4.0	1421	1477	3.94	5.0	1681	1712	1.84
4.5	1549	1589	2.58	6.0	1802	1895	5.16
5.0	1677	1686	0.54	7.0	1896	2071	9.23
5.5	1804	1763	−2.27	7.4	1977	2124	7.44
6.0	1855	1839	−0.86	–	–	–	–
6.4	1893	1879	−0.74	–	–	–	–

prediction results of the SVM model trained by three groups of training sample set are respectively shown in Tables 7.5, 7.6 and 7.7, where U_{50} is the experimental 50% discharge voltage given in [6–8], U_{50p} is the predicted result and σ is the relative error compared with the experimental data.

It can be seen that the relative errors of the predicted results are within 10%, except for only one sample. The maximum relative error e_{max} and the mean absolute

Table 7.8 SVM parameters and errors

Parameters or errors	Results A	Results B	Results C
C	415.8732	831.7470	702.3975
γ	0.1088	0.0583	0.0085
$e_{\max}$ (%)	10.81	8.44	9.23
e_{MAPE} (%)	6.84	4.19	3.46

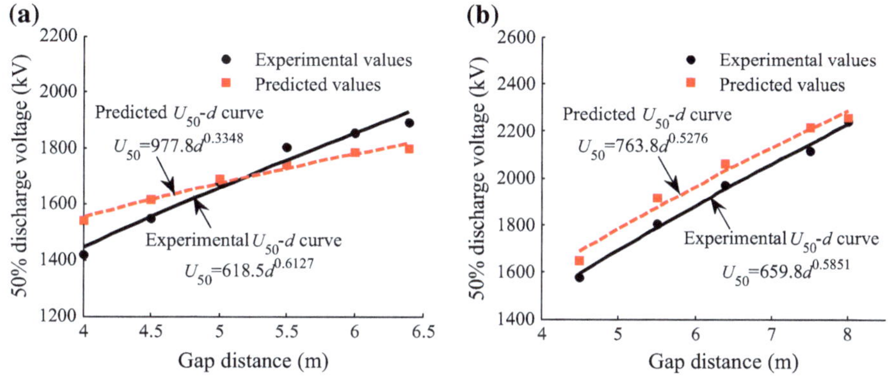

Fig. 7.14 Comparisons between the predicted and experimental U_{50} of transmission line-tower air gaps: **a** side-phase air gaps of 750 kV double-circuit transmission line, **b** middle-phase air gaps of UHV cup-type tower

percentage error e_{MAPE} are used to evaluate the overall prediction accuracy of the SVM model. The SVM parameters and the error results of the three predictions with different training sample sets are shown in Table 7.8. It can be seen that the e_{MAPE}s are 6.84, 4.19 and 3.46%, which are within the acceptable range in the view of engineering applications.

Taking the prediction case by training sample set B for example, the comparisons between the predicted and experimental U_{50}-d relations are shown in Fig. 7.14a, b, respectively for air gaps of 750 kV double-circuit tower and UHV cup-type tower. IEC 60071-2 [11] recommends the fitting formula between the positive switching impulse discharge voltage and the gap distance as $U_{50} = k \cdot 500d^{0.6}$, using the experimental results of positive rod-plane air gaps as a reference. Since the configurations of transmission line-tower gaps are quite different from rod-plane gaps, the exponent is also taken as the fitting parameter, thus to improve the fitting accuracy. Using $U_{50} = a \cdot d^b$ to fit the experimental and predicted results of 750 kV double-circuit tower and UHV cup-type tower, the fitting formulas are shown in Fig. 7.14. It can be seen that the predicted results can satisfy the accuracy requirements for engineering applications. Since the finite element models used for electric field calculation are simplified to a certain extent compared with the actual test configurations, the electric field features used for discharge voltage prediction may have some difference with the actual electric field distributions, thus may lead to certain errors inevitably.

From the above analysis, it can be seen that the U_{50} of EHV and UHV transmission line-tower air gaps can be predicted by the SVM model, trained by the experimental data of long rod-plane air gaps and those of an engineering gap configuration. This research demonstrates the feasibility of discharge voltage prediction for complicated engineering gaps. During the engineering practices of EHV and UHV transmission line constructions, numerous long air gap discharge test data have been accumulated. If these data can be used as training samples, the discharge voltages of new engineering air gaps can be predicted by the SVM model, which can effectively guide the external insulation design of power transmission and transformation projects, thus to save the required test work and test cost [12].

7.3 Discharge Voltage Prediction of Complex Gaps for Helicopter Live-Line Work

Live-line work is an important method for transmission line detection, maintenance and reconstruction. Helicopter live-line work (HLLW) can reduce the physical output of the line worker and improve the work efficiency, it has become an important measure for high voltage transmission line maintenance [13–15]. The two reported methods of HLLW are platform method and suspension or sling method, while the former has been used more commonly [16]. For the platform method, a worker sits on an operating platform electrically bonded to a helicopter, and the helicopter flies to the vicinity of the transmission line to position the worker, thus to perform the live-line work.

In order to guarantee safety, the required minimum gap distances between the helicopter and the live conductor, the grounded electrode during the live-line work should be studied. During the process of entering equipotential, the helicopter and the line worker are floating potential conductors in the electric field between the live conductor and the grounded electrode, and therefore a complex conductor-helicopter-ground air gap is formed. Currently, the minimum safe gap distance for live-line work, that is, the complex gap configuration corresponding to the minimum discharge voltage, is obtained mainly by gap discharge tests simulating the typical live-line work conditions. In this section, the switching impulse U_{50} of the complex gaps for helicopter live-line work are predicted by the SVM model. The predicted results are compared with the experimental data in references to validate the feasibility of the air insulation prediction model for discharge voltage prediction of engineering complex gaps.

7.3.1 Complex Gap Discharge Voltage Prediction Method

A complex gap with a floating object can be divided into two serial gaps. Taking a conductor-plane gap with a floating sphere for example, as shown in Fig. 7.15, it is composed of the gap 1 between the high voltage conductor and the floating sphere, and the gap 2 between the floating sphere and the grounding plane. The lengths of gap 1 and gap 2 are respectively d_1 and d_2, while the gap spacing between the conductor and the plane is d.

The complex gap discharge voltage prediction consists of the following steps:

1. First electric field calculation and feature extraction.

A 3D finite element model of the complex gap is established by a finite element analysis software. The high voltage conductor is applied a unit potential, the grounding electrode and the truncation boundary are applied zero potential, while the potential degree of freedom is coupled for the floating object. Then, the first electric field calculation is conducted. According to their gap geometries, the artificial discharge channel and the shortest path are defined respectively for gap 1 and gap 2. Their electric field features in the discharge channel and on the shortest path are extracted from the first electric field calculation results.

2. First discharge voltage prediction.

A prediction model is established by SVM, and it is trained by some experimental data of air gaps which have similar structures with gap 1 and gap 2. The discharge voltages of gap 1 and gap 2 are predicted by the SVM model. If the model outputs 1, it means that the electric potential difference between the high voltage conductor and the floating object, or that between the floating object and the grounding electrode are able to cause breakdown of the corresponding gap. The first discharge voltage

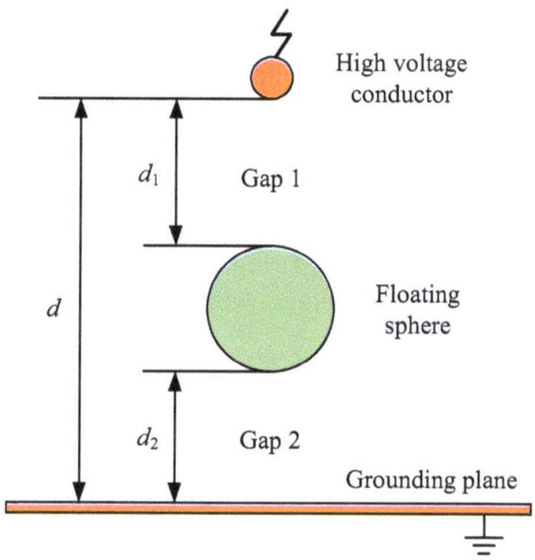

Fig. 7.15 Schematic diagram of a conductor-plane gap with a floating sphere

prediction results of gap 1 and gap 2 are respectively denoted as U_1 and U_2. If $U_1 < U_2$, it means that gap 1 will discharge firstly and U_1 is not sufficient for the breakdown of gap 2. Then the gap 1 is referred to as the primary gap, and the gap 2 is the secondary gap. Otherwise, it means that gap 2 will discharge firstly.

3. Second electric field calculation and feature extraction.

Taking the case of $U_1 < U_2$ for example, after gap 1 breakdown, the floating object can be considered to be equipotential with the high voltage conductor, while the arc voltage drop is neglected. According to the potential variation after gap 1 breakdown, the floating object and the high voltage conductor are applied the same potential for the second electric field calculation. The electric field features of gap 2 are extracted again from the FEM calculation results.

4. Second discharge voltage prediction.

The electric field features of gap 2 are input to the trained SVM model after normalization processing, and the discharge voltage of gap 2 is predicted secondary by the SVM model. Set the predicted result as U_2', if $U_2' > U_1$, it means that U_2' is sufficient for the breakdown of both gap 1 and gap 2. Then the discharge voltage of the complex gap is determined as U_2'. Otherwise, if $U_2' < U_1$, it means that gap 2 breakdown occurs soon after gap 1 breakdown under the applied voltage U_1. Then the final prediction result is U_1.

For the case of $U_1 > U_2$ in step 2, the floating object is equipotential with the grounding electrode, and the second electric field calculation is carried out for feature extraction of gap 1, so as to predict its discharge voltage again. The second prediction result of gap 1, set as U_1', is compared with the first prediction result of gap 2, that is, U_2. Similarly, the larger one is taken as the predicted complex gap discharge voltage.

The discharge voltage prediction procedure of the complex gap with a floating electrode is shown in Fig. 7.16. From the above-mentioned realization process, it can be seen that the first electric field calculation and discharge voltage prediction can be used to judge which gap will breakdown firstly and obtain its discharge voltage. By the second electric field calculation and discharge voltage prediction, the discharge voltage of the later breakdown gap can be obtained. By comparing the discharge voltages of the two gaps which breakdown successively, the whole discharge voltage prediction result of the complex air gap can be obtained. It should be noted that the above-mentioned predicted discharge voltage of gap 1 or gap 2 refers to the required potential applied on the high voltage electrode for discharge completion. Therefore, by comparing the twice discharge voltage prediction results, the complex gap discharge voltage, which makes two gaps breakdown, can be obtained.

7.3.2 Complex Gaps for Helicopter Live-Line Work

In order to determine the safe gap distance for helicopter live-line work on UHV AC transmission line, a test platform of helicopter live-line work was set up at the UHV AC test base in Wuhan, China. The relevant tests were carried out, including

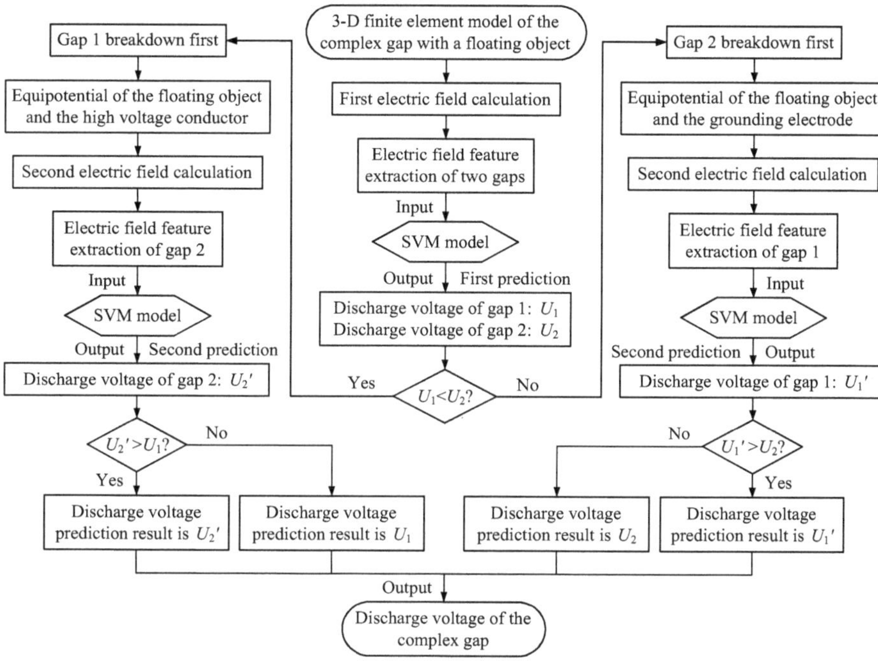

Fig. 7.16 Discharge voltage prediction procedure of the complex air gap

the switching impulse discharge tests of the complex gaps under typical phase-ground live-line work condition [13]. The test arrangement is shown in Fig. 7.17. This condition was used to simulate the process of entering equipotential, while the helicopter platform and the line worker are floating objects in the air gap between the energized transmission line and the overhead ground wire. As shown in Fig. 7.17, the complex gap consists of the gap 1 between the energized 8-bundled conductor and the line worker, and the gap 2 between the helicopter and the ground wire.

The simulated helicopter with a metal shell used in this experiment is full-scale made according to the size of the MD 500 helicopter [14, 15]. The simulated conductor is a 25 m long 8-bundled conductor, with a grading ring arranged at both ends, while the high-voltage lead wire of the impulse voltage generator is connected with the grading ring. The simulated ground wire is a 20 m long single conductor. The simulated line worker is made of aluminum alloy and fixed on the side of the test platform and the helicopter. The helicopter live-line work platform is above the insulation test stand with the height of 12 m. The top platform is a cuboid with the size of 3 m × 2.4 m × 0.2 m. The insulating supports consist of the upper and the lower parts, and the pedestal of the lower part consists of four cube stone piers with the side length of 1.2 m.

For the phase-ground discharge tests, the applied voltage is the positive 250/2500 μs switching impulse voltage. It should be noted that the main rotor blade

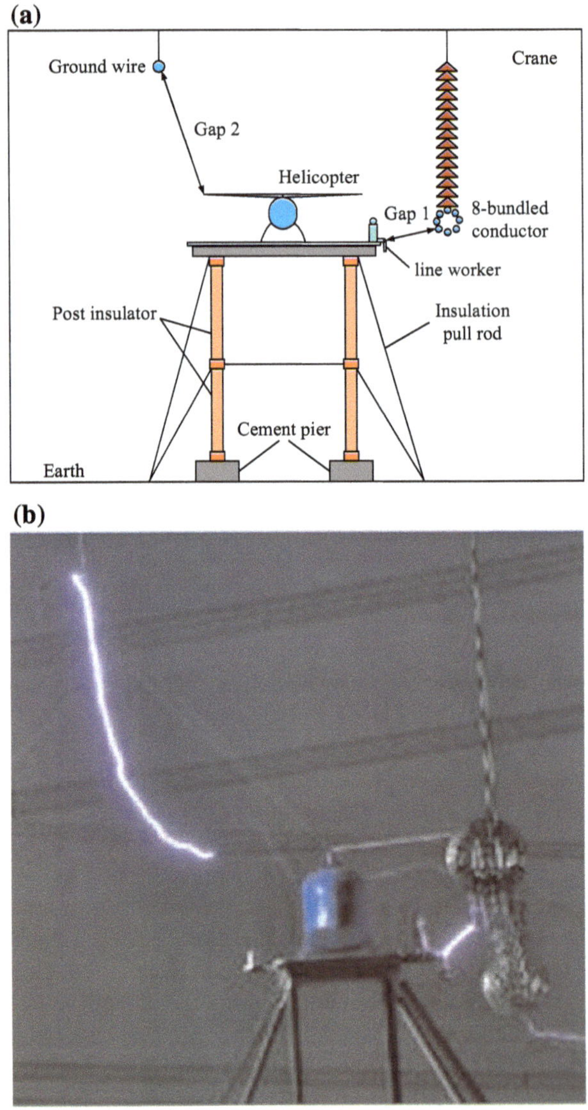

Fig. 7.17 Schematic diagram of complex gap for helicopter live-line work: **a** schematic diagram, **b** test arrangement

of the MD 500 helicopter used for actual live-line work is made of non-metallic composite light material, but the simulated helicopter in the outdoor test platform is made of metallic material. Therefore, the discharge channel can be formed between the main rotor blade and the ground wire. In order to study the discharge characteristics of the middle-phase complex gap for helicopter live-line work, the sum of the distance of gap 1 and gap 2 is kept as 7.5 m during the tests. For different test arrangements, the gap 1 and the gap 2 are respectively changed by moving the conductor and the ground wire using a crane to form different gap combinations. Finally, 6 test arrangements of the complex gap for helicopter live-line work are formed, while d_1/d_2 are respectively 0.5 m/7.0 m, 1.0 m/6.5 m, 1.6 m/5.9 m, 2.2 m/5.3 m, 2.8 m/4.7 m and 3.2 m/4.3 m. In this section, the 50% discharge voltages of the above complex gaps for helicopter live-line work are predicted by the SVM model, and the results are compared with the experimental values in [13].

7.3.3 Electric Field Calculation and Feature Extraction

According to the test configuration and the structure parameters of the outdoor test platform, helicopter, simulated 8-bundled conductor, ground wire and line worker, a three-dimensional electric field simulation model is established in SolidWorks software, as shown in Fig. 7.18. In order to decrease the calculation quantity, the simulation model is simplified appropriately in the finite element analysis.

The simulation model mainly contains the following parts:

1. Helicopter model.

The rotor blades are modelled based on the actual sizes and the test arrangement. The cabin of the helicopter is simplified as an ellipsoid, which is not modelled rigidly according to its actual appearance. The legs of the helicopter are ignored, since they

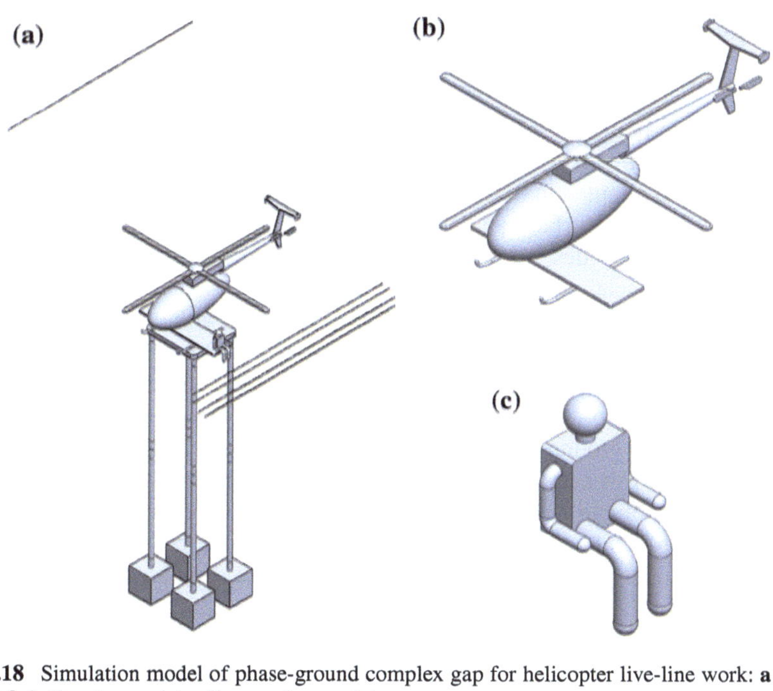

Fig. 7.18 Simulation model of phase-ground complex gap for helicopter live-line work: **a** whole model, **b** helicopter model, **c** line worker model

have little influence on electric field distributions in the discharge channels of gap 1 and gap 2. The undercarriage is with small radius of curvature, which may lead to electric field distortion, and therefore it is modelled as the actual shape. A 4.14 m long, 1.05 m wide operating platform is placed under the fuselage and connected with the undercarriage. The line worker sits on the outside of the operating platform when performing live work on the energized transmission lines. The helicopter model is shown in Fig. 7.18b.

2. Line worker model.

The shape and size of the line worker with a sitting posture in the experiments and simulations are based on the data reported in Chinese standard GB 10000-88 [17]. Accurate human body is very complex and the line worker model is simplified by a spherical head with the diameter of 22 cm, a 95 cm × 50 cm × 22 cm cuboid body with smoothed shoulders, and cylindrical arms and legs with the diameters of 8 and 12 cm. The line worker model is shown in Fig. 7.18c.

3. Models of the test stand, conductor and ground wire.

Since the influence of the insulation pull rods on the line worker and its nearby electric field distribution is negligible, they are neglected in the model. Meanwhile, the insulating ropes used to hang the simulated conductor and the ground wire, and the high voltage lead wire at the conductor end can also be neglected. The 8-bundled conductor and the ground wire are modelled according to the actual sizes used in 1000 kV transmission lines. The diameter of the subconductor is 30 mm and the bundle spacing is 400 mm [14].

4. Truncation boundary.

The truncated air boundary can be set as a rectangular solid or a semi-cylinder, while the latter is more effectively to reduce element quantity and save computation time, and therefore it is adopted in the FEM calculation here. For electric field calculation, the upper cylindrical surface and the earth are applied zero potential, while the vertical semicircular sides, which are perpendicular to the bundled conductor, are applied natural truncation boundary.

The simulation model was established to be as similar with the test configuration as possible. The complex gap discharge tests were reported in [13]. Gap 1 discharge happens between the conductor and the knee of the line worker, not between the conductor and the rotor blade, and gap 2 discharge happens between the helicopter rotor blade and the ground wire. Therefore, the discharge channel and the shortest path of gap 1 and gap 2 are defined in Fig. 7.19 for electric field feature extraction. The shortest path of gap 1 is the straight path from the end of the line worker's knee to its nearest sub-conductor, and that of gap 2 is the straight path from the tip of the rotor blade to the ground wire. The discharge channel is a rectangular area between two electrodes of the gap with the center of the shortest path.

For the first electric field calculation and feature extraction, the two discharge channels, operator and the tip of the main rotor blade are meshed more intensively. The conductor is applied 1 V, and the ground wire, earth and outer air boundary are

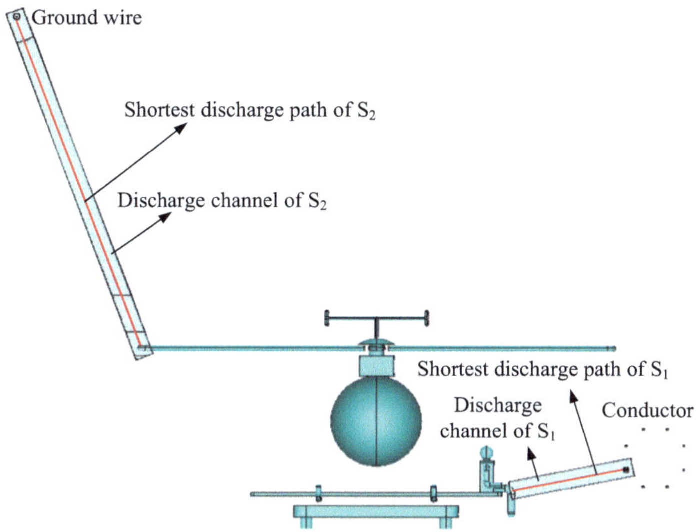

Fig. 7.19 Schematic diagram of the discharge channel and the shortest discharge path of phase-ground complex gap for helicopter live-line work

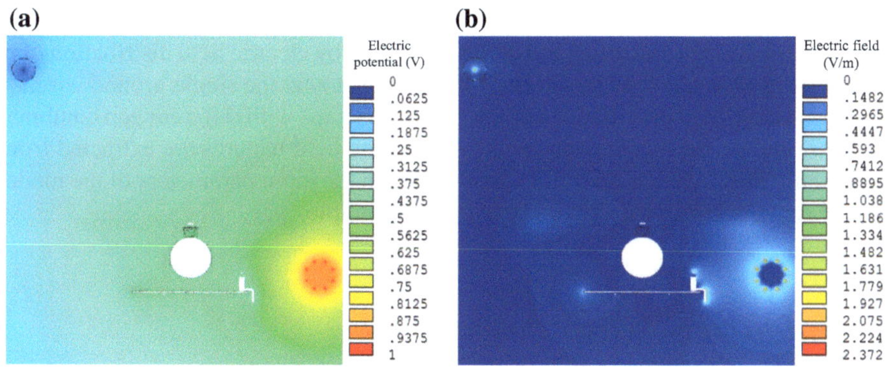

Fig. 7.20 First electric field calculation results of phase-ground complex gap for helicopter live-line work: **a** potential distribution, **b** electric field distribution

applied zero potential. The surface potential of the helicopter, platform, operator are coupled. Taking $d_1 = 1.6$ m, $d_2 = 5.9$ m for example, the potential and electric field distributions of the complex gap model for helicopter live-line work are shown in Fig. 7.20.

It can be seen from Fig. 7.20b that the electric field concentration appears on the front surface of the operator and the back end of the platform, and the electric field strength is relatively high between the operator and the bundled conductor. The electric field distributions along the shortest discharge path of the two gaps are shown in Fig. 7.21. It can be seen that the electric field distributions on the shortest discharge

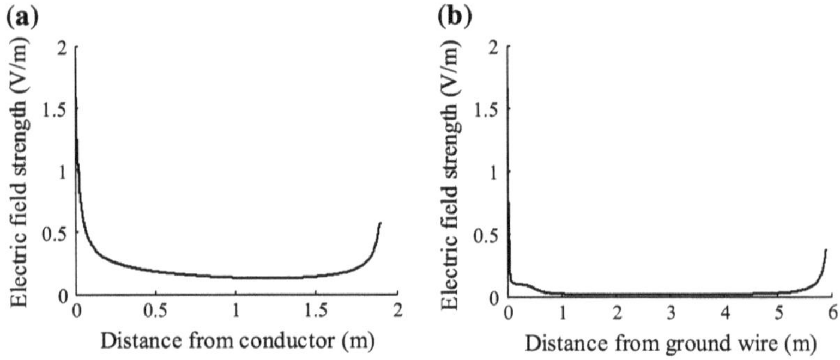

Fig. 7.21 Electric field distributions along the shortest discharge path of the complex gap for helicopter live-line work: **a** gap 1, **b** gap 2

path of gap 1 and gap 2 are both U-shape curves. The electric field strength is a bit high near the conductor and the ground wire, and the change law of the field strength reflects the electrode structure characteristics of gap 1 and gap 2. The electric field strength near the electrodes at two ends of gap 1 changes more gently than that of gap 2. The conductor is a multi-bundled structure, and the ground wire is a single conductor. The multi-bundled structure is equivalent to increasing the equivalent radius of the whole conductor, and therefore its nearby electric field distribution may be more uniform and changes more gently than that near the single ground wire.

The electric field distributions of the 6 complex gaps with different gap combinations are calculated successively, and their electric field features are extracted from the electric field calculation results, thus to obtain the input parameters of the model for the first discharge voltage prediction of the complex gap.

7.3.4 Discharge Voltage Prediction

7.3.4.1 Training Sample Set

Since the complex gap for helicopter live-line work is complicated and the structural characteristics of the two gaps are different, therefore the suitable training sample set can be selected respectively for discharge voltage prediction of the two gaps. The electrodes at two ends of gap 1 are respectively the bundled conductor and the operator's leg, and the electric field distribution along the shortest discharge path has the characteristics of rod-rod and sphere-plane air gaps. Besides, due to the complicated structure of the bundled conductor, the side-phase air gaps of UHV cup-type tower shown in Table 7.3 are also taken as the training samples, together with the rod-rod and sphere-plane air gaps. The electrodes at two ends of gap 2 are respectively a single ground wire and the main rotor blade of the helicopter, and the electric field distribution has the characteristics of rod-rod and rod-plane air gaps.

Table 7.9 50% discharge voltages of rod-rod and sphere-plane gaps under standard switching impulse

Rod-rod gaps		Sphere-plane gaps				
d (m)	U_{50} (kV)	*d* (m)	U_{50} (kV)			
			$D = 25$ cm	$D = 45$ cm	$D = 75$ cm	$D = 95$ cm
2	989	2	800	864	1126	–
3	1253	3	1020	1045	1222	1504
5	1627	4	1154	1186	1331	1550
10.5	2701	5	1286	1303	1412	1585

Since its structure is relatively simple and without the bundled conductor, the training sample set for discharge voltage prediction of gap 2 contains only the experimental data of rod-rod and rod-plane air gaps. The sample data of rod-plane gaps are shown in Table 7.4, and the sample data of rod-rod and sphere-plane gaps are cited from [18] and [19], as shown in Table 7.9, where *D* represents the sphere diameter of the sphere-plane gap.

According to the electrode structures and test arrangements of rod-rod and sphere-plane gaps given in [18] and [19], their finite element models are established for electric field calculation and feature extraction. Together with the rod-plane gas and the side-phase air gaps of UHV cup-type tower, these gaps constitute the training sample set for discharge voltage prediction of the phase-ground complex gaps for helicopter live-line work.

7.3.4.2 First Prediction of the Discharge Voltage

For the gap 1 and gap 2 of the complex gap for helicopter live-line work, the corresponding training sample set are used to train the SVM model. The optimal parameters of the SVM model for gap 1 are $C = 68.59$, $\gamma = 0.406$, and those for gap 2 are $C = 194.01$, $\gamma = 0.812$. According to the discharge voltage prediction procedure of the complex gap, the electric field features extracted from the first electric field calculation results are taken as the input parameters. The discharge voltages of gap 1 and gap 2 are predicted respectively by the SVM model, and the predicted results are shown in Fig. 7.22.

It can be seen from Fig. 7.22 that the discharge voltage prediction results of gap 1 and gap 2 increase with the length of gap 1. For every gap combination, the predicted discharge voltage of gap 1 is less than that of gap 2. When the length of gap 1 increases, the floating potential on the operator will decrease, and therefore it requires larger applied voltage for the breakdown of gap 2. Although the length of the gap 2 becomes shorter, the influence is less than that of the decrease of the floating potential. It should be noted that the discharge voltage prediction value refers to the required voltage applied on the conductor for the gap breakdown, without considering whether another gap will discharge or not. According to the discharge voltage prediction procedure

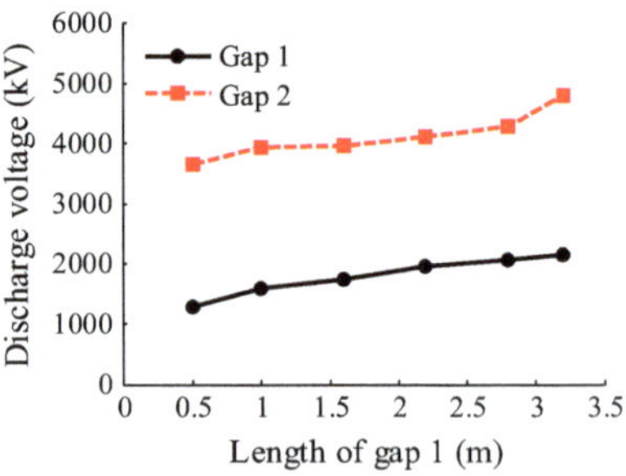

Fig. 7.22 First prediction results of the discharge voltage of gap 1 and gap 2

of the complex gap shown in Fig. 7.16, since the predicted discharge voltage of gap 1 is less than that of gap 2 under each gap distance, it can be considered that gap 1 will breakdown at first. The arc voltage is much smaller than the breakdown voltage, therefore it can be neglected, and the operator can be considered as equipotential with the high voltage conductor. Under this condition, the second electric field calculation is conducted for feature extraction and the second prediction of the discharge voltage.

7.3.4.3 Second Prediction of the Discharge Voltage

For the second electric field calculation, the bundled conductor, helicopter, test platform and operator are applied 1 V, and the ground wire, earth and outer air boundary are applied zero potential. Taking the complex gap of $d_1 = 1.6$ m, $d_2 = 5.9$ m for example, the potential and electric field distributions are shown in Fig. 7.23. It can be seen that the second electric field calculation result is different from that of the first calculation. The helicopter live-line work platform and the operator are also with high potential after gap 1 breakdown. Due to the edge effect of the conductor, the high electric field strength is mainly concentrated at the operator's feet, the end of the test platform and the outer tip of the main rotor blade.

For the second discharge voltage prediction of the complex gaps, the applied training sample set and SVM model are the same with the first prediction. Taking the electric field features extracted from the second electric field calculation results as input parameters, the discharge voltages of gap 2 with 6 gap combinations are predicted by the SVM model. Under different gap combinations, the first discharge voltage prediction values of gap 1 and the second discharge voltage prediction values of gap 2 are shown in Fig. 7.24. It can be seen that $U_2' > U_1$ when $d_1 = 0.5$, 1.0, 1.6, and 2.2 m, and $U_1 > U_2'$ when $d_1 = 2.8$ and 3.2 m. The former ones indicate that gap 2 breakdown requires a larger applied voltage U_2' after gap 1 breakdown. The latter ones mean that gap 2 will discharge immediately after gap 1 breakdown under U_1.

The final predicted discharge voltage of the complex gap is the larger one in the two predictions. The predicted 50% discharge voltage of the phase-ground complex gap for helicopter live-line work can be obtained from Fig. 7.24, as shown in Table 7.10,

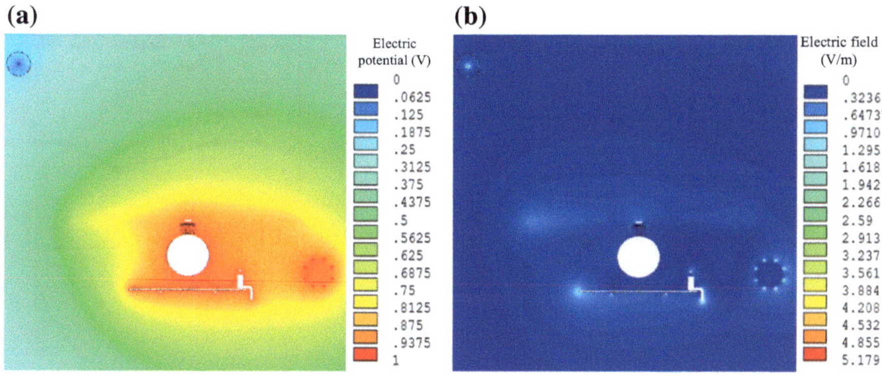

Fig. 7.23 Second electric field calculation results of phase-ground complex gap for helicopter live-line work: **a** potential distribution, **b** electric field distribution

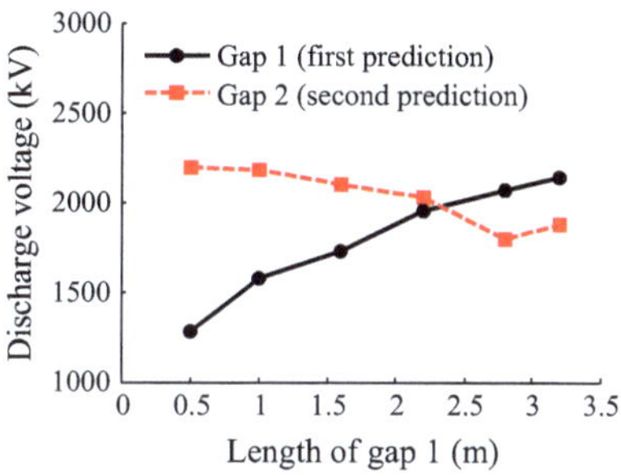

Fig. 7.24 Comparison between the first and the second discharge voltage prediction results of complex gap

Table 7.10 Discharge voltage prediction results of the phase-ground complex gap for helicopter live-line work

d_1 (m)	d_2 (m)	U_{50t} (kV)	U_{50p} (kV)	σ (%)
0.5	7.0	2080	2201	5.8
1.0	6.5	2075	2184	5.3
1.6	5.9	2000	2103	5.2
2.2	5.3	1950	2031	4.2
2.8	4.7	2065	2068	0.1
3.2	4.3	2070	2138	3.3

where U_{50t} and U_{50p} are respectively the experimental and predicted 50% discharge voltage, and σ is their relative error.

It can be seen from Table 7.10 that the maximum relative error between the predicted and experimental U_{50} is 5.8%, and the e_{MAPE} of the 6 test samples is 4.0%. The errors are within the acceptable range for engineering application. The comparison between the predicted and experimental discharge voltages of different gap combinations are shown in Fig. 7.25. It can be seen that there is a U-shape curve for both predicted and experimental results, and the minimum discharge voltage of

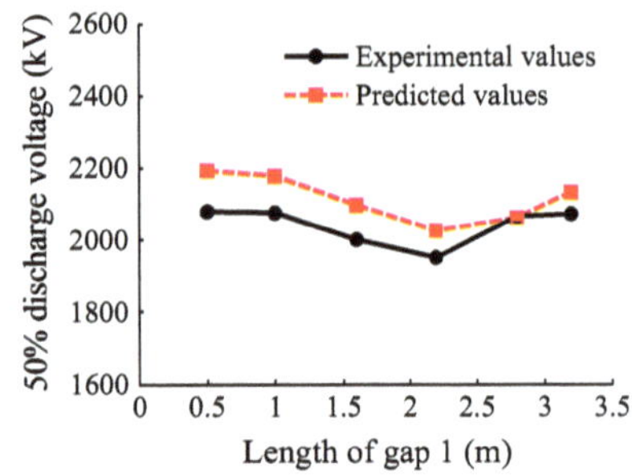

Fig. 7.25 Comparison between the predicted and experimental discharge voltages of complex gaps for helicopter live-line work

the two curves appears at the same gap distance, that is, $d_1 = 2.2$ m, which further verifies the rationality of the SVM prediction results.

7.4 Brief Summary

In this chapter, the impulse discharge voltages of 3 kinds of engineering gaps are predicted by the air insulation prediction model, including the parallel gaps for transmission line insulator string, the transmission line-tower air gaps and the complex gaps for helicopter live-line work. The contents can be summarized as follows:

1. Taking 6 test data of rod-rod gaps as training samples, the lightning impulse 50% discharge voltages of 8 parallel gaps for transmission line insulator string can be predicted. Compared with the experimental data corrected by 3 altitude correction formulas, the e_{MAPE} of the predicted results are 3.03%, 3.55% and 4.71%, respectively. The errors are within the acceptable range for engineering application. The discharge voltage prediction by SVM model can provide reference for engineering design of parallel gaps.
2. Taking the experimental switching impulse U_{50} of the long rod-plane air gap and an engineering gap configuration as training samples, the U_{50} of the middle-phase air gaps on the 750 kV double-circuit tower, the bottom-phase V-string air gaps of UHV compact tower, and the side-phase air gaps of the UHV cup-type tower are predicted by the SVM model. The e_{MAPE}s of the predicted results are 6.84%, 4.19% and 3.46%, respectively. The proposed air insulation prediction model is able to predict the discharge voltages of transmission line-tower air gaps, which can provide guidance for external insulation design of power transmission projects.
3. Taking typical air gaps and the engineering air gaps as training samples, the 50% discharge voltages of the phase-ground complex gaps for helicopter live-line work are predicted by the SVM model. The errors are within acceptable range for engineering application, while the e_{MAPE} of the predicted results is 4.0%. The predicted U_{50}-d curve has the same trend with the experimental results, which verifies the validity of the air insulation prediction model for discharge voltage prediction of complex gaps with engineering configurations.

References

1. Sima WX, Ye X, Tan W et al (2012) Lightning insulating co-ordination between insulator string and parallel gap device of 220 kV transmission line at high altitude area. Proc CSEE 32(10):168–176
2. Zhang Z (2013) Study on insulation coordination and arc guiding performance of ring shape parallel gap lightning protection devices for 220 kV insulator strings. Dissertation, Chongqing University
3. IEC 60071-1 (2006) Insulation co-ordination—part 1: definitions, principles and rules
4. Wan QF, Chen Y, Huo F et al (2011) High altitude correction method for air insulation gap of UHV and EHV lines. China Patent 200710169012.7, 6 July 2011
5. Liao YL, Li RH, Li XJ et al (2012) Experimental research on typical air gap test voltage correction. Proc CSEE 32(28):171–176
6. Chen Y, Meng G, Xie L et al (2008) Research on air-gap discharge characteristics of 750 kV one tower double-circuit transmission line. High Volt Eng 34(10):2118–2123
7. Huo F (2012) Study on insulation characteristics and electric field distribution of long air-gaps for UHV power transmission line. Dissertation, Wuhan University
8. Huo F, Hu W, Xu T et al (2011) Air-gaps flashover characteristics for 1000 kV AC compact tower. High Volt Eng 37(8):1850–1856
9. Paris L (1967) Influence of air gap characteristics on line-to-ground switching surge strength. IEEE Trans Power Appar Syst 86(8):936–947
10. Thione L, Pigini A, Allen NL et al (1992) Guidelines for the evaluation of the dielectric strength of external insulation. CIGRE Brochure, Paris, France
11. IEC 60071-2 (1996) Insulation co-ordination—part 2: application guide
12. Qiu ZB, Ruan JJ, Jin Q et al (2015) Switching impulse discharge voltage prediction of EHV and UHV transmission lines-tower air gaps by a support vector classifier. IET Gener Transm Distrib 12(15):3711–3717
13. Du Y, Peng Y, Liu T et al (2015) Experimental research for safe gap distance of helicopter live working with platform method on UHV AC transmission line. High Volt Eng 41(4):1292–1298
14. Liao CB, Ruan JJ, Liu C et al (2016) Helicopter live-line work on 1000-kV UHV transmission lines. IEEE Trans Power Deliv 31(3):982–989
15. Liu C (2017) Study on approaching path and safety of helicopter live-line work on UHV AC transmission lines. Dissertation, Wuhan University
16. IEEE Task Force 15.07.05.05 (2000) Recommended practices for helicopter bonding procedures for live-line work. IEEE Trans Power Deliv 15(1):333–349
17. GB 10000-88 (1988) Human dimensions of Chinese adults
18. Xie SJ, He HX, Xiang NW et al (2012) Experimental study on the discharge process of rod-rod air gap under switching impulse voltage. High Volt Eng 38(8):2083–2090
19. Wang X (2010) The comparison of critical radius of rod-plane gap at different altitudes and research on altitude correction. Dissertation, China Electric Power Research Institute
20. Qiu ZB, Ruan JJ, Liu C et al (2018) Discharge voltage prediction of complex gaps for helicopter live-line work: an approach and its application. Electr Power Syst Res 164:139–148

MIX
Papier aus verantwortungsvollen Quellen
Paper from responsible sources
FSC® C105338

If you have any concerns about our products,
you can contact us on
ProductSafety@springernature.com

In case Publisher is established outside the EU,
the EU authorized representative is:
Springer Nature Customer Service Center GmbH
Europaplatz 3, 69115 Heidelberg, Germany

Printed by Libri Plureos GmbH
in Hamburg, Germany